Energy Resources

by Laura McDonald

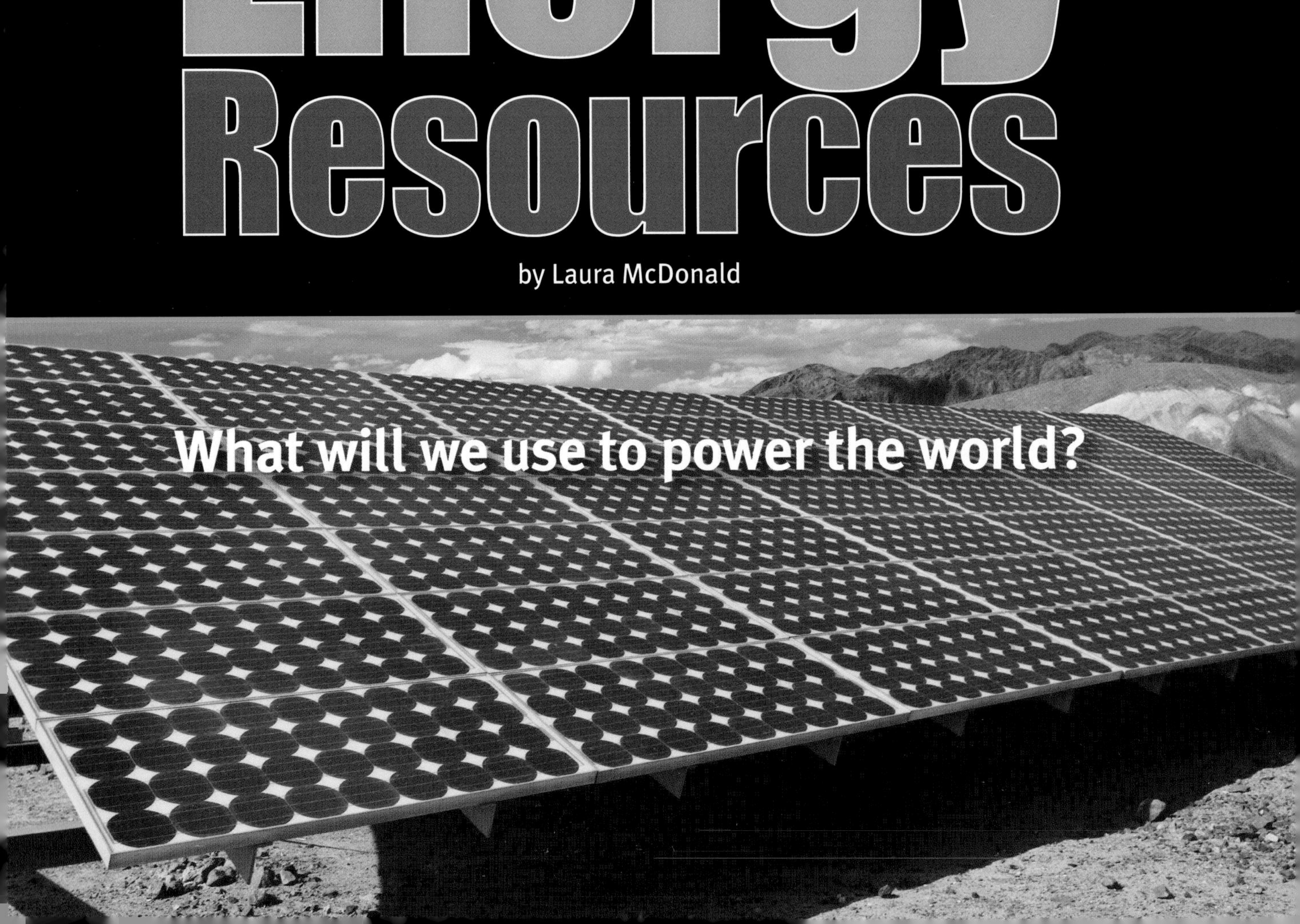

What will we use to power the world?

TABLE OF CONTENTS

What are the advantages and limitations of our planet's energy resources?

INTRODUCTION

The Biggest Dam

CHINA IS A NATION OF 1.3 BILLION PEOPLE. All those people need energy to drive cars and power their homes. They also need energy to run businesses and factories. Where will China find enough energy?

For years, China has relied on coal as a major energy resource. However, coal mining is dangerous and burning coal produces a lot of pollution. China also imports oil from other countries. But burning oil also causes pollution. The energy of moving water, on the other hand, is clean and available from China's many rivers.

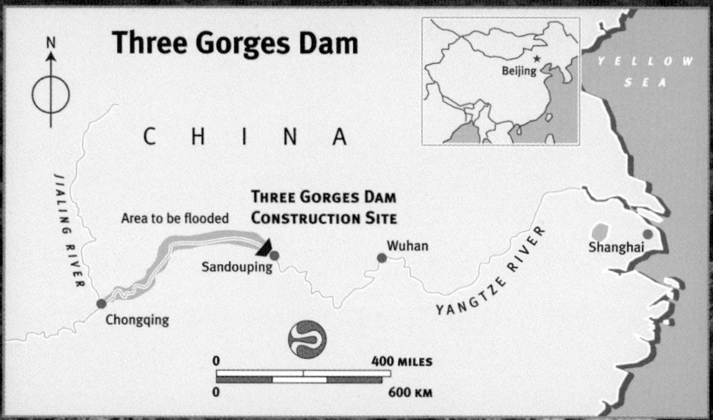

The Yangtze (YANG-see) is one of those mighty rivers. The river carries needed water from the mountains to the city of Shanghai. However, the Yangtze also brings terrible floods. Records show that the Yangtze flooded 1,000 times in the last 2,000 years. In 1931, a flood killed 140,000 people. The same flood forced 12 million people to move off their land.

The Chinese have long dreamed of controlling the Yangtze River. Their boldest idea was the Three Gorges Project. This plan would build the largest dam in the world, 185 meters (606 feet) high. Water pouring through the dam would produce 20 billion watts of electricity. This is enough to power the state of New Jersey.

The Three Gorges Project would flood a large area while producing electricity and controlling the Yangtze River.

in the World

The Three Gorges proposal created an uproar in China and around the world. Millions of people live near the Yangtze. Thousands of ancient towns dot the banks. How many would be covered by water trapped behind the dam?

Many rare plants and animals also make their homes along the Yangtze. For example, the Chinese river dolphin lives only in the Yangtze River. Scientists doubted that the dolphin could survive the building of the dam.

China faced a great challenge. Leaders had to balance the need for clean energy and flood control with the needs of local people and wildlife.

Coal-burning power plants cause much of the smog in China.

CHAPTER 1

What Are Energy Resources?

OUR MODERN WORLD IS POWERED BY ENERGY. CARS, TELEVISIONS, TOASTERS, AND FACTORIES ALL REQUIRE ENERGY TO FUNCTION. WHAT EXACTLY IS ENERGY? WHAT SOURCES OF ENERGY CAN PEOPLE USE?

▲ People use energy for many different purposes.

What Is Energy?

Energy is one of the most basic features of the universe. Light, heat, sound, and electricity are all forms of energy. Energy is the ability to do work. In the picture below, energy moves the cars down the street. Lamps need energy to make light. The trees need energy to grow. A lot of energy went into building this city, as well.

Stored energy is called **potential energy**. Food, firewood, and gasoline all contain potential energy. **Kinetic energy** is energy of movement. It is the energy an object has because of the motion of its mass. Flowing water and blowing air have kinetic energy. A cyclist riding down a hill gains kinetic energy. As she increases her speed, her kinetic energy also increases. The faster something moves, the more kinetic energy it has. Energy can be converted from one form to another. Each time energy changes form, some of it turns into heat.

Essential Vocabulary

- energy page 7
- energy resource page 8
- kinetic energy page 7
- nonrenewable resource page 8
- potential energy page 7
- renewable resource page 8

The Root of the Meaning

The word **energy** comes from the Greek word *energos*, meaning "work."

How do people choose among and use energy resources?

Energy Sources

Energy comes from many different sources. An energy source used to meet the needs of people is called an **energy resource**. You can think of an energy resource as the raw material from which energy is produced.

The sun is our most important source of energy. Sunlight, or solar energy, is a **renewable resource** that nature replaces in a short period of time. The energy of wind, waves, and running water also comes indirectly from solar energy.

Plants convert the kinetic energy of light into the potential energy of food. The food provides energy for growth and reproduction. Animals gain this energy when they eat plants. People use plant and animal materials, or biomass, as sources of energy, too. Some of the energy in the remains of ancient living things is preserved deep in the ground. Coal, natural gas, and petroleum are all found in the ground. These fuels are called **nonrenewable resources** because they cannot be replaced.

The heat inside Earth is an energy resource that does not come from the sun. Another nonsolar energy resource is nuclear energy. Nuclear energy is the potential energy stored in atoms.

▼ People use special panels called solar cells to absorb solar energy.

▼ This wind farm uses hundreds of windmills to harness wind energy.

renewable

People use many different energy resources.

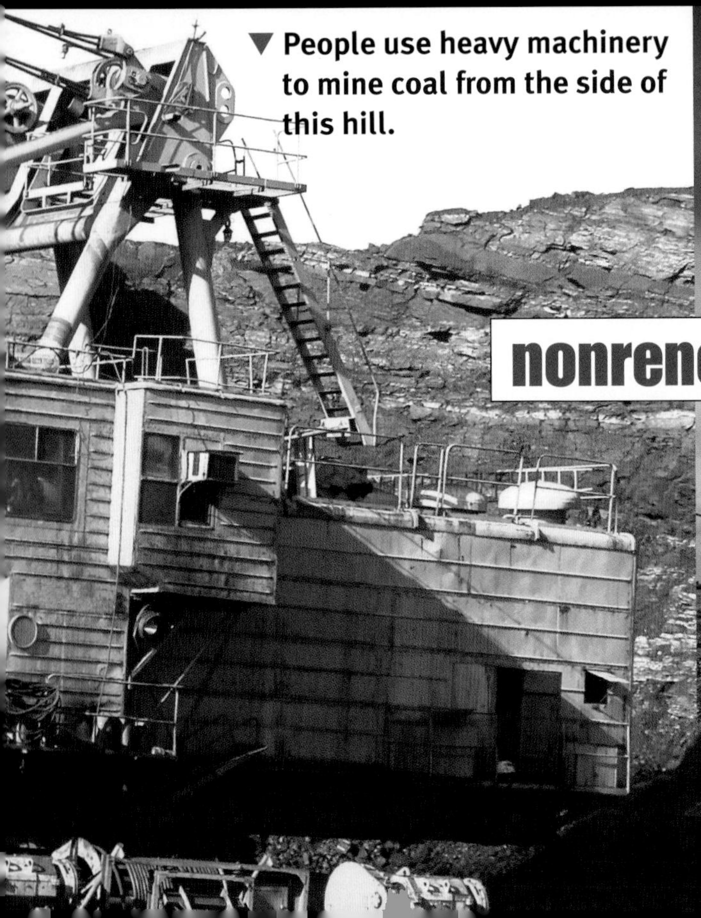

▼ People use heavy machinery to mine coal from the side of this hill.

nonrenewable

▲ This oil rig drills petroleum, or crude oil, from Earth's crust.

Using Energy Resources

Energy resources are all around us. However, it takes some work to make the resources useful. First, people must collect, or extract, the energy resources. People build dams to collect moving water. They build pumps to extract oil, and build solar cells to capture sunlight.

Next, pipelines, trucks, ships, and trains must transport these resources. The resources are brought to places where they will be processed, or refined. Then processing converts energy resources into useful fuels. Petroleum, or crude oil, starts out as a thick, gooey liquid. A refinery separates petroleum into different fuels.

CHECKPOINT

Reread

Pages 10 and 11 describe the steps needed to make use of the energy in energy resources. Reread these pages and share your findings with your classmates.

These fuels include gasoline, kerosene, and heating oil. Nuclear fuel is made from uranium. This element is mined from underground and purified. Processing can convert biomass and fossil fuels into clean-burning hydrogen gas.

Moving water is stored behind a dam so that its energy can be used.

Converting Energy

Finally, people must convert the potential energy in these resources to useful work. People use technology to convert energy. For example, car engines burn gasoline or ethanol. Potential energy in the fuel becomes the kinetic energy of a moving car.

Power plants also extract energy from these resources to make electricity. The energy of moving water, wind, or burning fuel turns a giant fan called a turbine. The turbine spins a generator that produces an electric current. The electricity can be stored in batteries. It can also be transported in power lines to homes and businesses.

All energy resources require the use of energy. It takes energy to power each step from collecting natural resources to using their energy. The energy that goes into processing and transporting energy resources is lost as heat to the surrounding water and air.

FROM ENERGY RESOURCE TO USEFUL WORK

Choosing Among Energy Resources

How do people choose which energy resources to use? People generally use the cheapest energy resource available. When the price of one resource gets too high, people switch to a less expensive resource.

Historically, people in Europe used wood for heat. As the European population increased, wood became more and more scarce. By the 1800s, the price of wood rose higher than the price of coal. Coal replaced wood as the most used energy resource. The current interest in "alternative" fuels is related to the rising cost of petroleum-based gasoline and diesel.

The cost of using an energy resource goes beyond money. Energy choices also affect national safety, people's health and happiness, and the environment. Many nations prefer energy resources found in their own country. People also want clean energy resources that won't pollute the air and water. Pollution causes serious problems for all living things.

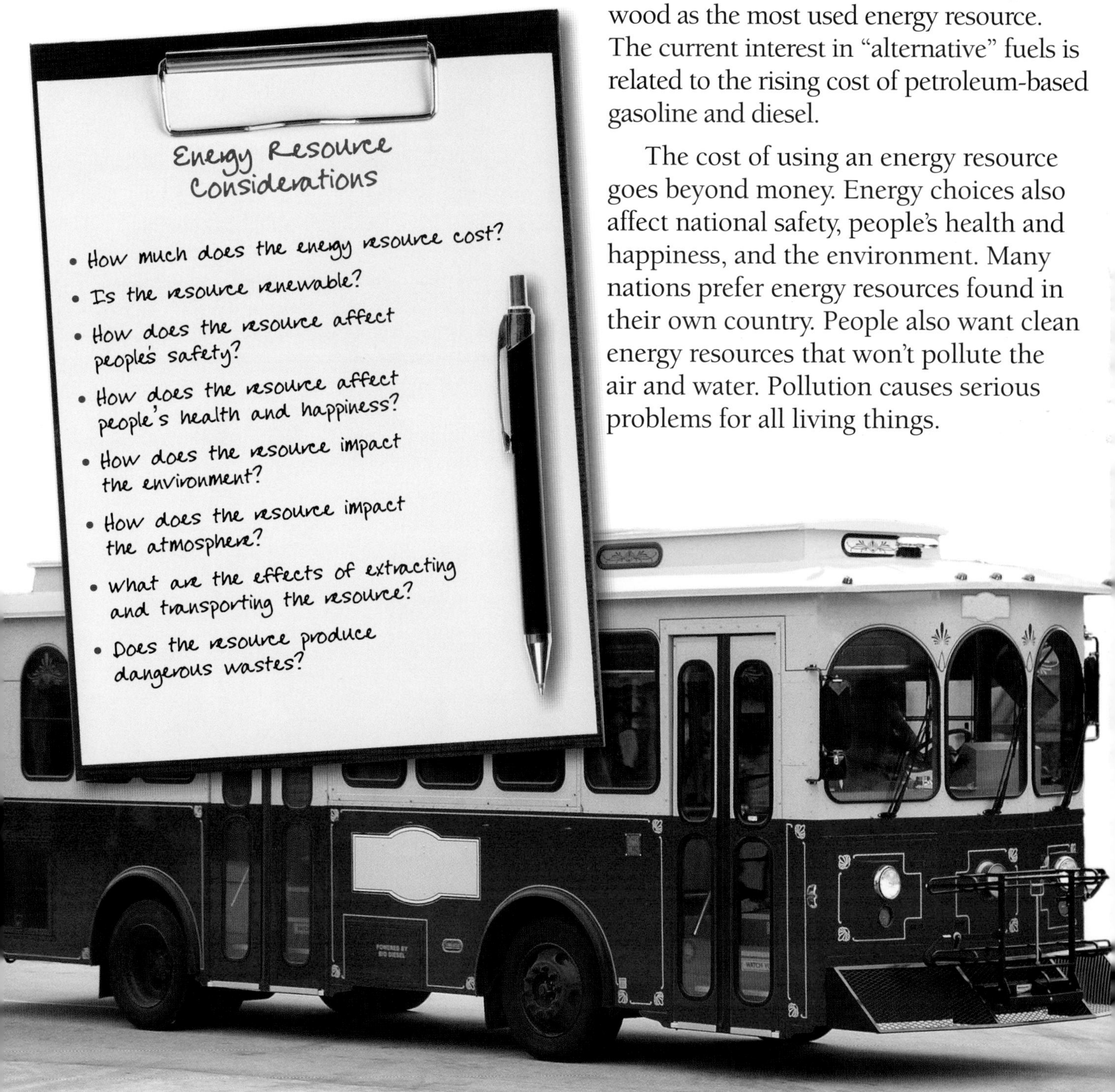
Energy Resource Considerations
• How much does the energy resource cost?
• Is the resource renewable?
• How does the resource affect people's safety?
• How does the resource affect people's health and happiness?
• How does the resource impact the environment?
• How does the resource impact the atmosphere?
• What are the effects of extracting and transporting the resource?
• Does the resource produce dangerous wastes?

▲ When the costs of one energy resource increase, people start to explore new resources.

Pollution contributes to the warming of Earth's atmosphere. Plants, animals, and people will all have to adjust to new weather patterns, increased storms, and growing deserts. The change in climate results partly from carbon dioxide released by burning fossil fuels.

Scientists around the world are exploring ways to reduce global climate change. Some solutions include switching to renewable energy resources.

Increasing efficiency can also reduce the carbon dioxide released into the air. Efficiency is a measure of the useful work produced by a certain amount of energy. Increasing efficiency means using less energy to provide the same level of energy service. If a car can go twice as far on a tank of gas, it is twice as efficient. The car also produces half as much carbon dioxide.

CHECKPOINT

Talk It Over

The cost of using gasoline is much higher than the price people pay at the pump. Hold a class debate about whether the price paid for energy resources should include all the costs, from extraction to disposal of wastes.

In order to make wise choices about energy resources, you need to know all the effects of using them. The next three chapters explore the advantages and disadvantages of each of Earth's energy resources.

▼ Polar bears depend on sea ice, which is shrinking due to global climate change.

Hands-On Science

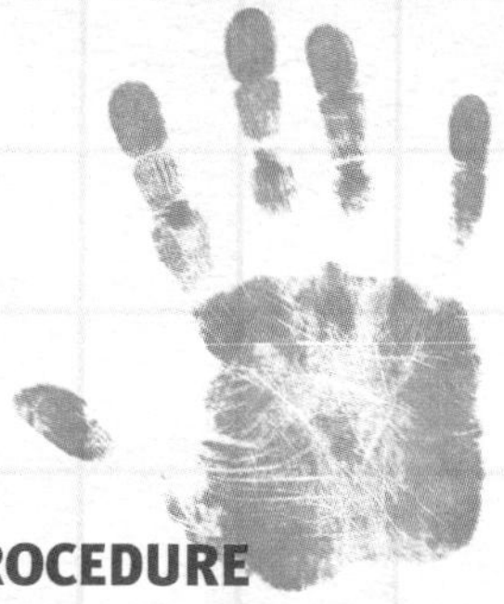

Feel the Heat

Do you need a new lightbulb? In the store, you can choose between regular incandescent lights and compact fluorescent lights (CFLs). Which type of lightbulb uses electricity more efficiently?

TIME REQUIRED

20 minutes

MATERIALS NEEDED

- desk or table lamp
- compact fluorescent light (CFL) and incandescent lightbulb with the same light output/wattage
- thermometer
- ring stand
- centimeter ruler

SAFETY CONSIDERATIONS

Incandescent lightbulbs get very hot—do not touch the bulb after the experiment. Turn off the lamp before screwing in or removing a lightbulb.

PROCEDURE

1. Make sure the lamp is turned off. Screw in the CFL bulb. Turn on the lamp and observe the light produced.
2. Use a ring stand to hold the thermometer 15 cm (6 in) away from the glowing light. Record the temperature on the thermometer. After 10 minutes, record the temperature again.
3. Turn off the lamp and allow the bulb and thermometer to cool for 3 minutes. Remove the bulb.
4. Repeat steps 1 and 2 with the incandescent bulb. Turn off the lamp.

ANALYSIS

1. Compare the heat and light produced by each lightbulb.
2. Which lightbulb is more efficient at converting electricity to light?
3. Is it possible to create a lightbulb that produces only light and no heat? Explain your answer.

DATA TABLE

LIGHT TYPE	DESCRIBE THE LIGHT PRODUCED	TEMPERATURE MEASUREMENTS	CHANGE IN TEMPERATURE
Compact fluorescent light (CFL)		Start: Finish:	
Incandescent light		Start: Finish:	

SUMMING UP

- Energy resources provide people with the energy they need for modern life.
- People must collect, extract, transport, and process most energy resources before their energy can be used.
- Each energy resource offers advantages but also comes with limitations.
- People and governments consider many factors before they choose which energy resources to use.

Putting It All Together

Choose one of the activities below.

1. A magazine article claims to have found a perfect, free energy resource. Write a letter to the editor of the magazine responding to this claim.
2. Walk around your home and make a list of all the appliances and machines that require energy to work. Include the energy source for each item on your list. Compare your list with a friend's list.
3. Reread pages 12 and 13, and then predict what would happen if the price of petroleum became extremely high. Write a paragraph explaining how your life might be different after people adjusted to the high cost of petroleum.
4. You read on page 8 that sunlight causes the movement of wind. Research the energy conversions that lead to the creation of wind. Make a diagram and share your findings with your class.

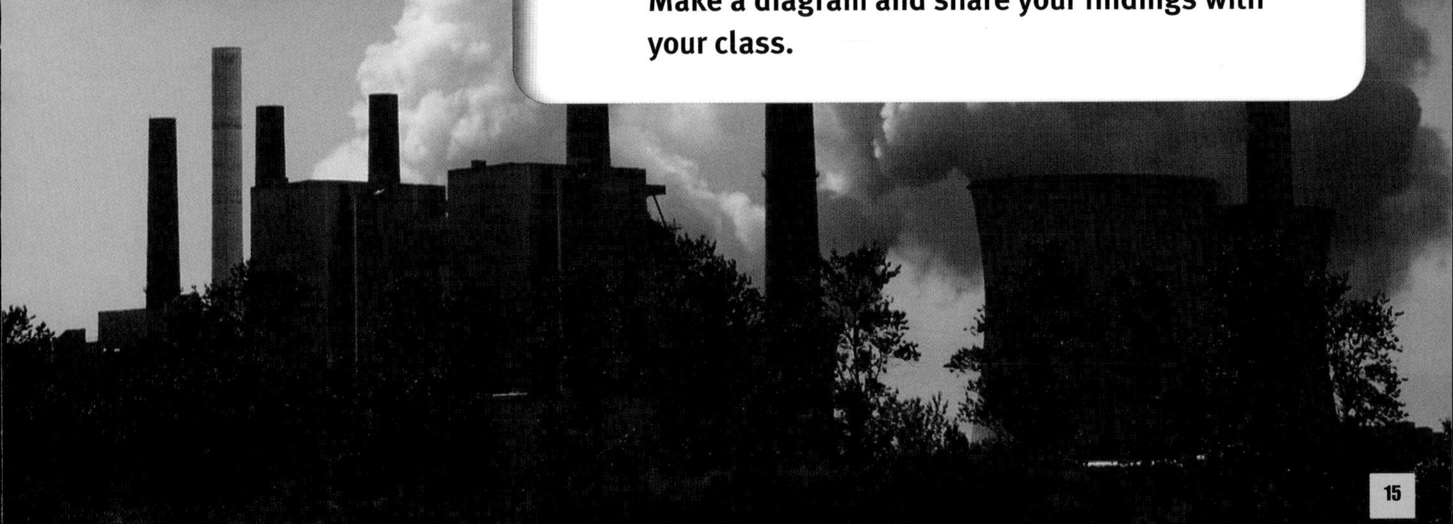

CHAPTER 2

Energy from Fossil Fuels

HAVE YOU EVER SEEN A LUMP OF COAL, A SAMPLE OF CRUDE OIL, OR A CONTAINER OF NATURAL GAS? EVEN IF YOU HAVEN'T, THESE FOSSIL FUELS IMPACT YOUR LIFE. AMERICANS USE FOSSIL FUELS TO SATISFY 85% OF THEIR ENERGY NEEDS.

A **fossil fuel** is an energy resource that is formed inside Earth from ancient plant or animal remains. The biomass of the remains changed slowly over millions of years to form coal, natural gas, or petroleum. The diagrams on page 18 show the formation of each type of fossil fuel. Fossil fuels are nonrenewable resources because they take such a long time to form.

Fossil fuels include solid coal, gaseous natural gas, and liquid petroleum.

Essential Vocabulary

- **fossil fuel** page 16

What are the advantages and disadvantages of energy from fossil fuels?

Fossil fuels contain a lot of chemical potential (stored) energy. Burning fossil fuels converts this energy to heat energy. This heat energy can be used to warm buildings, generate electricity, or power factories. Fossil fuels are very important to modern society. At the same time, fossil fuels have several important limitations.

CHECKPOINT

Read More About It

Read more about the ancient forests and swamps that became the coal deposits of today. How were these forests like the forests of today? How were they different?

How Coal Formed

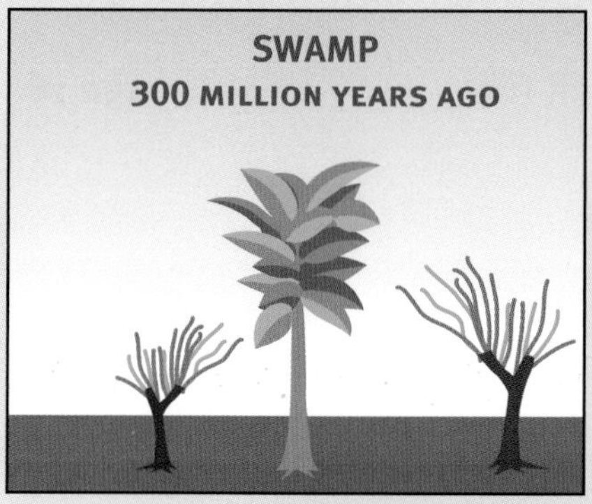

Before the dinosaurs, many giant plants died in swamps.

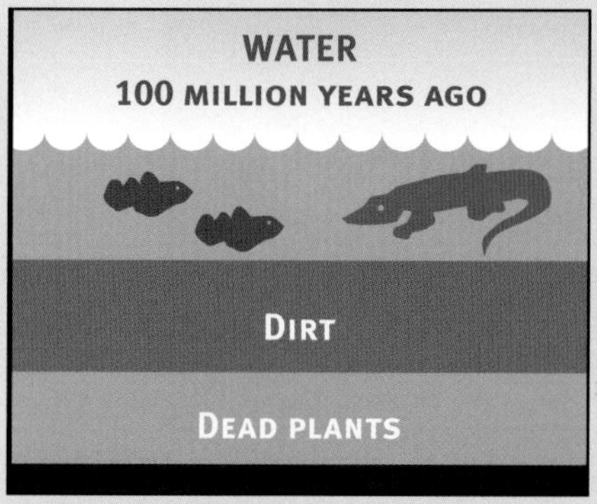

Over millions of years, the plants were buried under water and dirt.

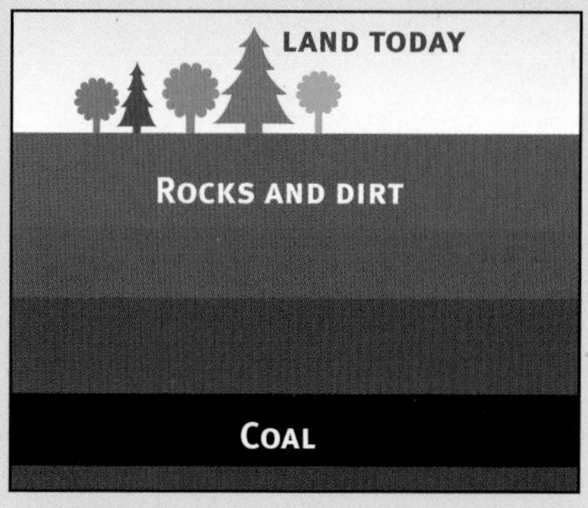

Heat and pressure turned the dead plants into coal.

How Petroleum and Natural Gas Formed

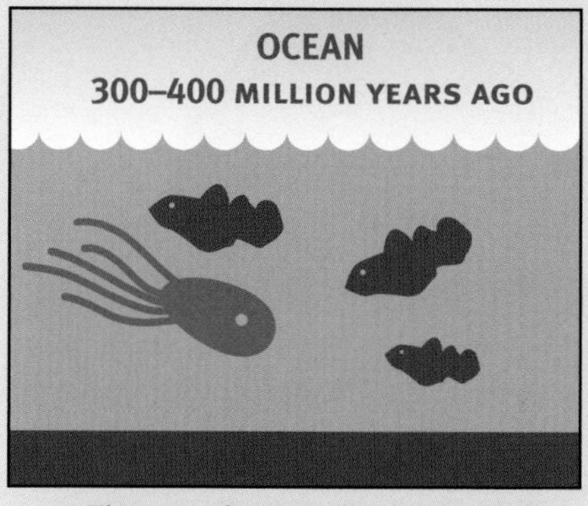

Tiny sea plants and animals died and were buried on the ocean floor. Over time, they were covered by layers of silt and sand.

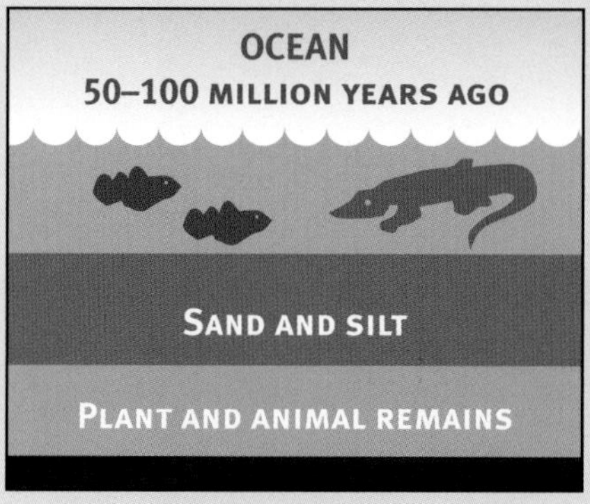

Over millions of years, the remains were buried deeper and deeper. The enormous heat and pressure turned them into oil and gas.

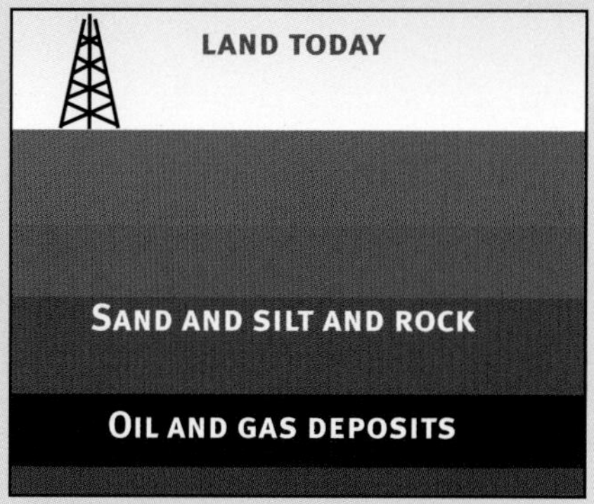

Today, we drill down through layers of sand, silt, and rock to reach the rock formations that contain oil and gas deposits.

Coal

As an energy resource, coal has many advantages. Coal is common in the United States. Twenty-six of the fifty states have coal deposits. Kentucky, Pennsylvania, Texas, Wyoming, and West Virginia mine the most coal. There is enough coal left to supply the country's needs for over a century. Once mined, coal is easy to transport and ready to burn. Coal costs less than most other energy resources. Because of these advantages, coal supplies Americans with 50% of their electricity and 22% of their overall energy needs.

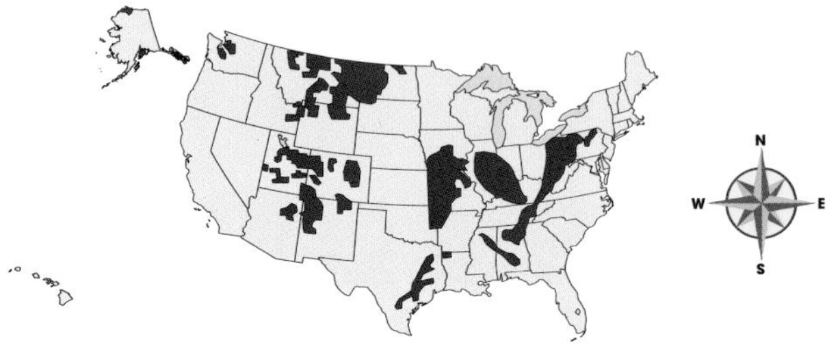

■ U.S. Coal-Producing Regions

Coal use also has serious disadvantages. Mining accidents and coal-related pollution lead to many thousands of deaths each year around the world. Burning coal is the worst source of air pollution and contributes to global climate change. Businesses, scientists, and the government are working together on clean coal technology. The goal of clean coal is to reduce the environmental impact of coal burning by purifying the coal and capturing carbon dioxide before it reaches the atmosphere.

How Coal Is Used

▼ Mining brings coal up to the surface where it can be used.

EVERYDAY SCIENCE

Natural gas has no odor. Processors add a strong-smelling chemical to alert people in case of a natural gas leak.

▲ The search for fossil fuels begins with geologists. These scientists study Earth's structure and processes. This pumping station pumps natural gas up from deposits found beneath the ocean floor.

Natural Gas

Natural gas is another plentiful and inexpensive energy resource. Burning natural gas converts its potential chemical energy to heat. Natural gas heats buildings, powers factories, and produces electricity. Some new vehicles run on natural gas fuel. Others run on hydrogen produced from natural gas.

Huge deposits of natural gas lie under the ground and in coastal waters. Natural gas even bubbles out of swamps and marshes. Natural gas is composed mostly of the colorless gas methane. Methane burns easily and cleanly. Methane produces no smoke or ash.

Despite its advantages, natural gas use also creates some problems. Burning natural gas produces carbon dioxide. To combat global climate change, some natural gas facilities pump carbon dioxide back underground. Tragic natural gas explosions have taken many lives over the years. However, natural gas is much cleaner and safer than coal production and use.

How Natural Gas Is Used

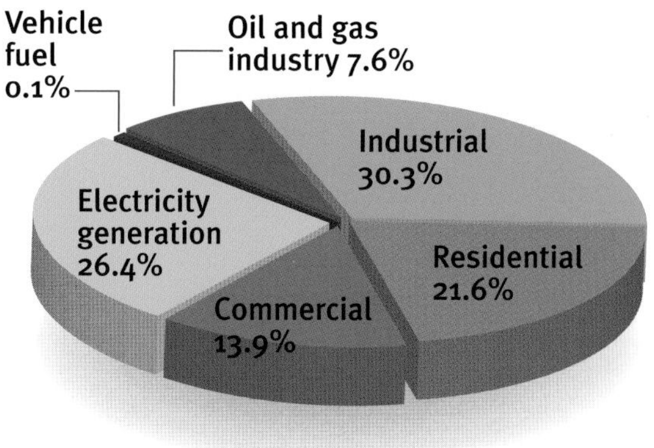

Petroleum

People around the world use more petroleum than any other energy resource. In its natural form, petroleum, or crude oil, is a thick, smelly, dark liquid. Pumps bring petroleum up from the ground or seafloor. Then ships and pipelines transport the crude oil to a refinery.

Crude oil contains thousands of different chemicals. This mixture is refined, or separated, to make different products. These products include gasoline, kerosene, plastic ingredients, tar, asphalt, fertilizers, food additives, and many others. Power plants also burn petroleum to produce electricity. Petroleum is very useful!

The petroleum deposits under the United States are not large enough to meet the country's needs. Texas, Alaska, California, Louisiana, and Oklahoma produce the most petroleum. Petroleum from the Middle East, South America, and other regions supplies the rest of the country's needs.

Experts disagree about the total amount of petroleum remaining in the world. This uncertainty is one reason why the prices of petroleum products rise and fall so often. However, it is safe to say that the world's reserves of petroleum are shrinking every year.

Products Made from Petroleum

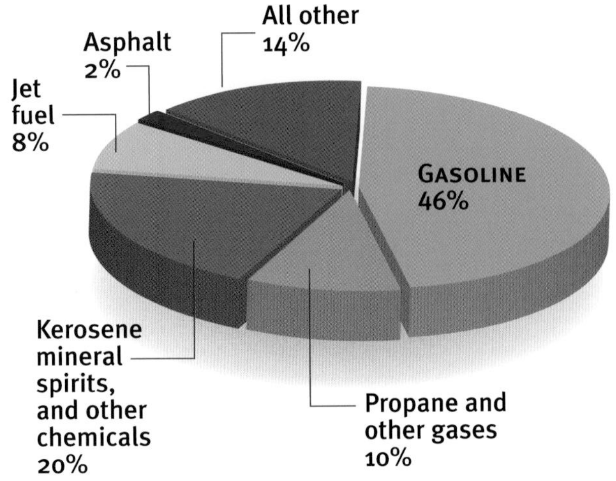

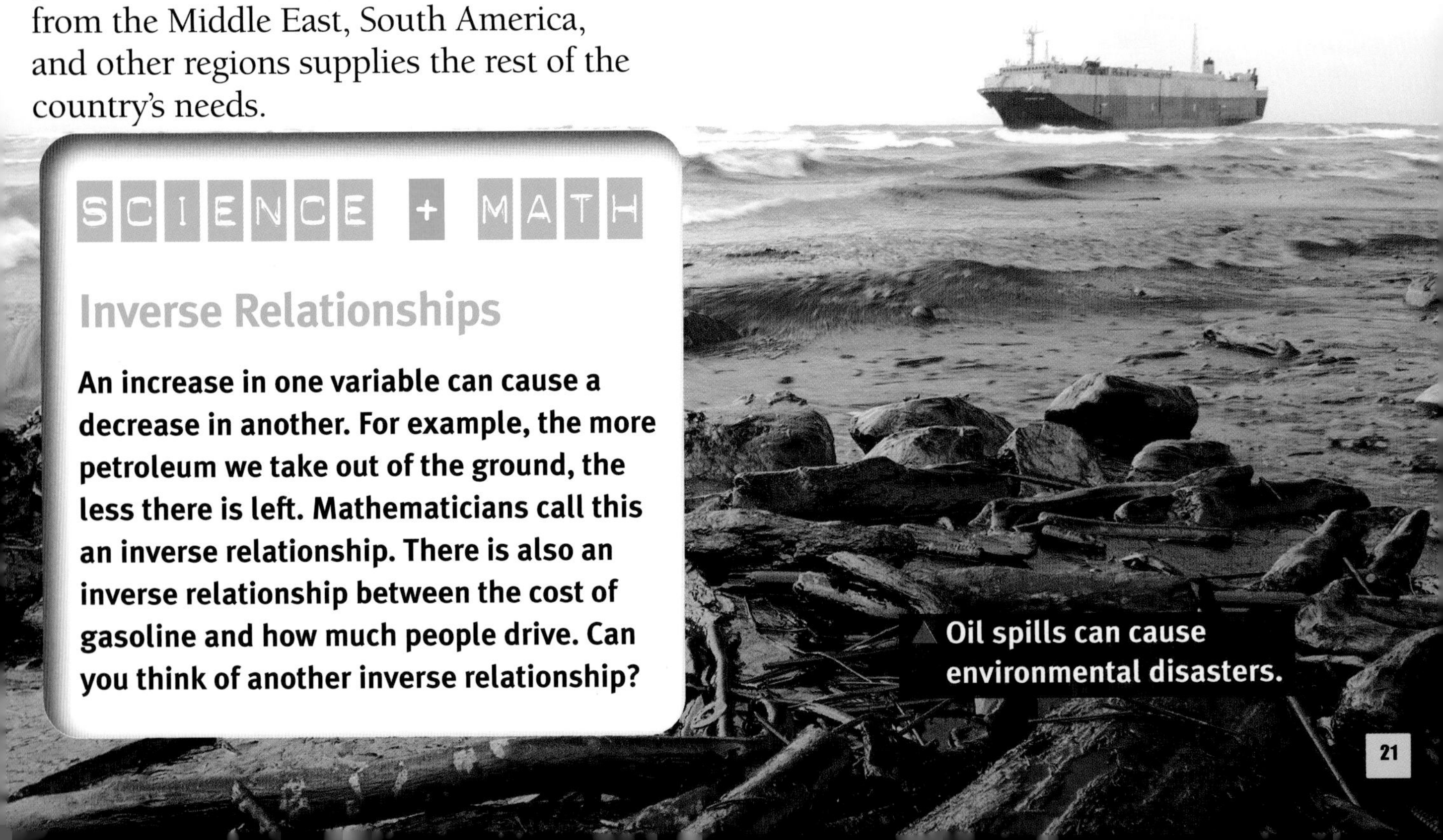

Oil spills can cause environmental disasters.

SCIENCE + MATH

Inverse Relationships

An increase in one variable can cause a decrease in another. For example, the more petroleum we take out of the ground, the less there is left. Mathematicians call this an inverse relationship. There is also an inverse relationship between the cost of gasoline and how much people drive. Can you think of another inverse relationship?

Petroleum also has its disadvantages. The environmental costs of petroleum use can be very high. Oil-rich areas are environmentally sensitive. New technologies have helped reduce the damages of drilling, but drilling still disrupts land and ocean habitats. Burning petroleum products produces air pollution, including carbon dioxide. Ships and pipelines carrying crude oil can leak or spill oil into the ocean. In the worst accidents, millions of liters of petroleum coat the water with a thick layer of oily gunk. Fish, seabirds, small animals, and plants die in large numbers.

Many crude oil deposits are located in politically unstable countries. The United States spends a lot of money and human resources to protect its petroleum supply.

The limited supply and environmental hazards of petroleum are forcing the world to look for alternatives. Finding a mix of energy resources to replace petroleum is one of this century's greatest challenges.

Advantages and Disadvantages of Fossil Fuels

ENERGY RESOURCE	ADVANTAGES	DISADVANTAGES
 Coal	• Large supply available in the United States • One of the least expensive energy resources • Easily transported by trains	• Nonrenewable resource • Mining damages environment • Mining accidents result in many deaths • Can produce air and water pollution
 Natural Gas	• Clean fuel • Requires little processing • Large supply available in the United States	• Nonrenewable resource • Contributes to global climate change by creating carbon dioxide • Highly flammable • Costly to transport
 Petroleum	• Contains many useful fuels • Easily transported liquid • Relatively inexpensive to produce	• Nonrenewable resource • Worldwide supply shrinking rapidly • Extraction, transportation, and burning cause environmental damage • Contributes to global climate change • Many deposits in politically unstable countries

SUMMING UP

- Fossil fuels form from the remains of ancient plants and animals.
- Petroleum, coal, and natural gas are extracted from mines and wells underground or under the sea.
- People burn these nonrenewable energy resources to power transportation, generate electricity, and produce heat.
- Modern society relies heavily on nonrenewable resources because they are inexpensive, convenient energy resources.
- However, supplies of fossil fuels are limited and their use can endanger human life and damage the environment.

Putting It All Together

Choose one of the activities below.

1. You hear a caller on a radio show suggest that drilling for oil in Alaska would solve the need for more petroleum. How would you respond to this caller? Write a paragraph explaining your response. Compare your response to a classmate's.

2. The coal industry is trying to reduce air and water pollution by creating clean coal technology. Read more about clean coal technology in a library or on the Internet. Show some of the new developments on a poster to share with your class.

3. Reread page 21 about the uses of petroleum. For one day, record all the times you use or observe items made from or powered by petroleum products.

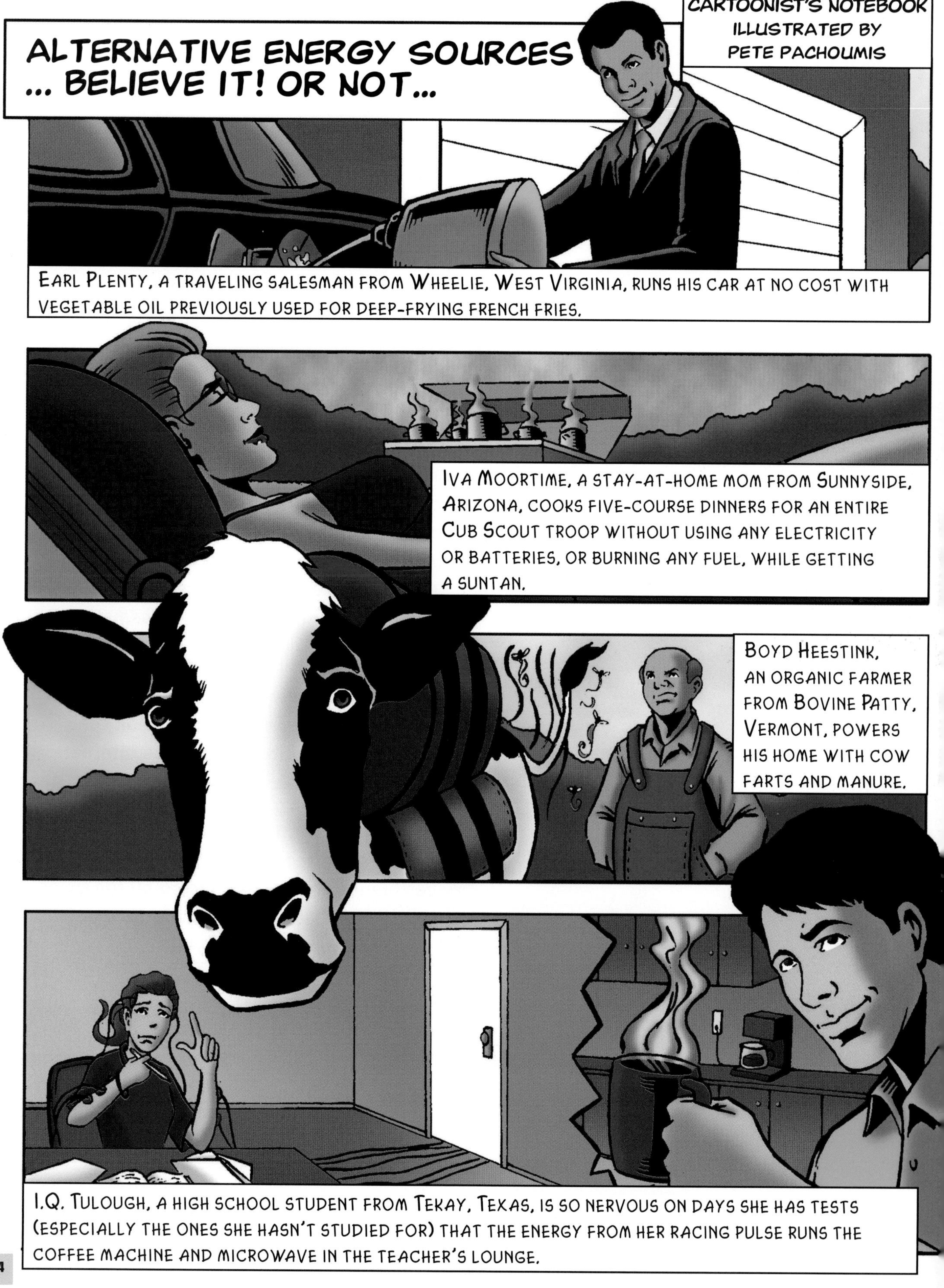
CARTOONIST'S NOTEBOOK
ILLUSTRATED BY
PETE PACHOUMIS
ALTERNATIVE ENERGY SOURCES ... BELIEVE IT! OR NOT...
EARL PLENTY, A TRAVELING SALESMAN FROM WHEELIE, WEST VIRGINIA, RUNS HIS CAR AT NO COST WITH VEGETABLE OIL PREVIOUSLY USED FOR DEEP-FRYING FRENCH FRIES.
IVA MOORTIME, A STAY-AT-HOME MOM FROM SUNNYSIDE, ARIZONA, COOKS FIVE-COURSE DINNERS FOR AN ENTIRE CUB SCOUT TROOP WITHOUT USING ANY ELECTRICITY OR BATTERIES, OR BURNING ANY FUEL, WHILE GETTING A SUNTAN.
BOYD HEESTINK, AN ORGANIC FARMER FROM BOVINE PATTY, VERMONT, POWERS HIS HOME WITH COW FARTS AND MANURE.
I.Q. TULOUGH, A HIGH SCHOOL STUDENT FROM TEKAY, TEXAS, IS SO NERVOUS ON DAYS SHE HAS TESTS (ESPECIALLY THE ONES SHE HASN'T STUDIED FOR) THAT THE ENERGY FROM HER RACING PULSE RUNS THE COFFEE MACHINE AND MICROWAVE IN THE TEACHER'S LOUNGE.

BELIEVE IT: ANY CAR WITH A DIESEL ENGINE CAN BE MODIFIED INTO A "GREASE CAR" THAT RUNS ON WVO — WASTE VEGETABLE OIL — THROWN OUT FROM RESTAURANTS. THE CAR WILL GET THE SAME GAS MILEAGE AND WON'T ADD EXTRA CARBON DIOXIDE TO THE ATMOSPHERE.
TRUE: SOLAR COOKERS USE THE SUN AS THEIR SOLE SOURCE OF ENERGY. THERE IS NO COST TO RUN THEM, NOR TO THE ENVIRONMENT. MANY OF THE DOZENS OF TYPES OF SOLAR COOKERS CAN BE BUILT FROM EVERYDAY MATERIALS IN JUST A FEW HOURS.
STINKY BUT TRUE: COW WASTE CONTAINS METHANE, A COLORLESS, THOUGH ALAS, NOT ODORLESS GAS. THIS GAS CAN BE CONVERTED TO ELECTRICITY.
WELL... THIS ONE ISN'T TRUE — YET!
MANY ALTERNATIVE ENERGY SOURCES MAY SEEM CRAZY AT FIRST, BUT SOME OF THESE IDEAS ACTUALLY WORK.
CAN YOU THINK OF OTHER IDEAS THAT PEOPLE THOUGHT WERE WACKY AT FIRST, BUT THEN CHANGED HISTORY?

CHAPTER 3

Nuclear Energy

THE CORE OF AN ATOM, CALLED A NUCLEUS, IS MADE OF POSITIVELY CHARGED PROTONS, AND NEUTRONS WHICH HAVE NO CHARGE. YOU WOULD THINK THAT BECAUSE THE NUCLEUS CONTAINS MANY POSITIVE CHARGES THAT THEY WOULD REPEL EACH OTHER AND IT WOULD BE IMPOSSIBLE TO KEEP A NUCLEUS INTACT. YOU WOULD BE CORRECT! HOWEVER, THERE IS A FORCE, CALLED THE STRONG FORCE, THAT, OVER VERY SHORT DISTANCES, IS GREATER THAN THE FORCE OF REPULSION BETWEEN THE PROTONS. THE STRONG FORCE HOLDS EVERYTHING IN THE NUCLEUS TOGETHER.

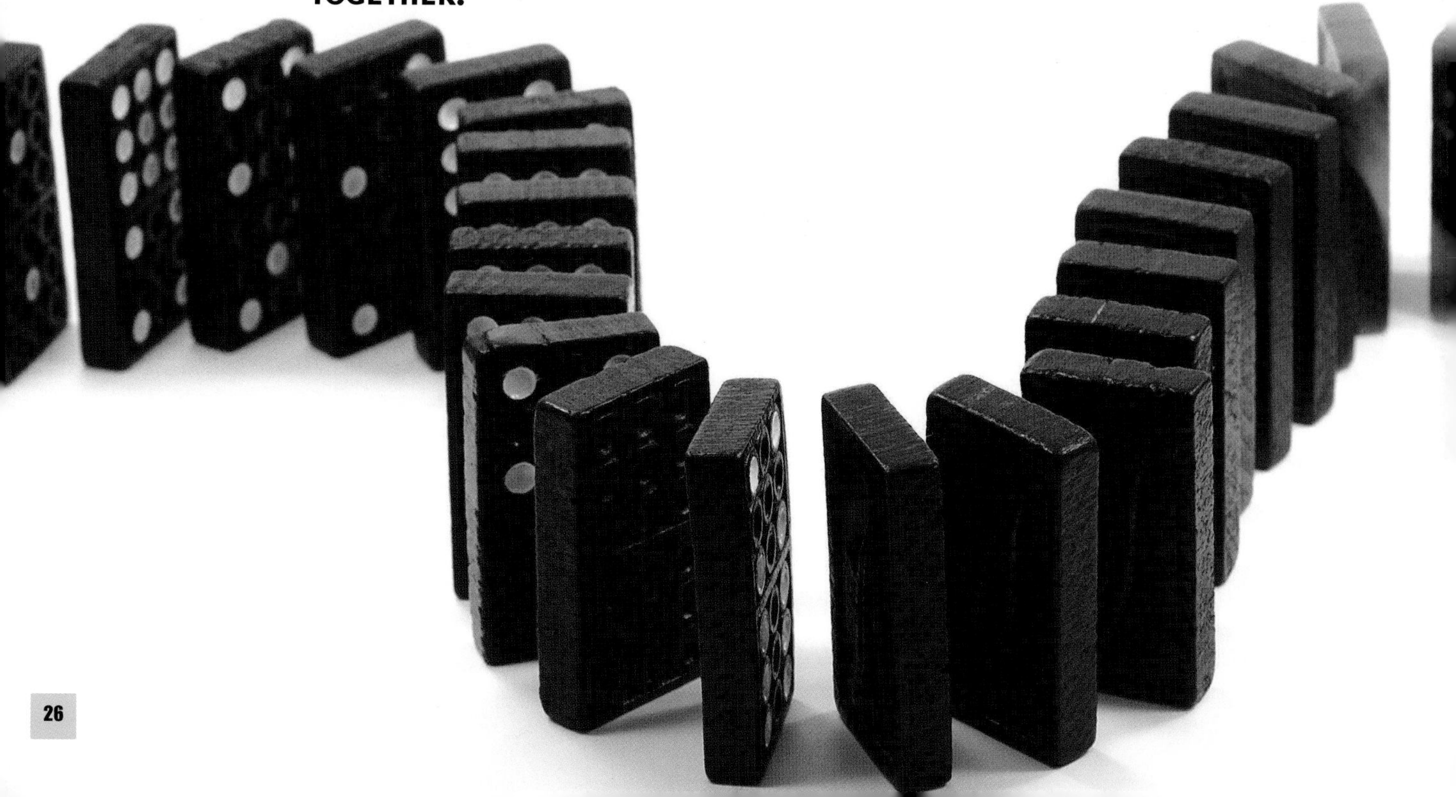

Using Nuclear Energy

Some types of atomic nuclei are **radioactive**. This means they spontaneously decay, or change, into other types of nuclei. As each radioactive nucleus decays, it releases energy, or radiation. Some nuclei may even undergo **fission.** Fission happens when a nucleus splits into two smaller nuclei and one or more other particles. This is where our story of nuclear energy begins.

Uranium-235 is radioactive. The 235 is the sum of the protons and neutrons in its nucleus. This type of uranium is commonly found in rocks all over the world.

Uranium-235 nuclei spontaneously split to form the nuclei of two different atoms, as well as two or three neutrons. These neutrons can be absorbed by other uranium-235 atoms, causing them to become unstable and to also undergo fission. As more and more neutrons are released, more and more uranium atoms are split. This is called a chain reaction. Unless there is something to absorb some of the free neutrons, the splitting of uranium continues and BOOM! The sum of all the energy that is released as each atom was split produces a tremendous explosion.

What are the advantages and disadvantages of nuclear energy?

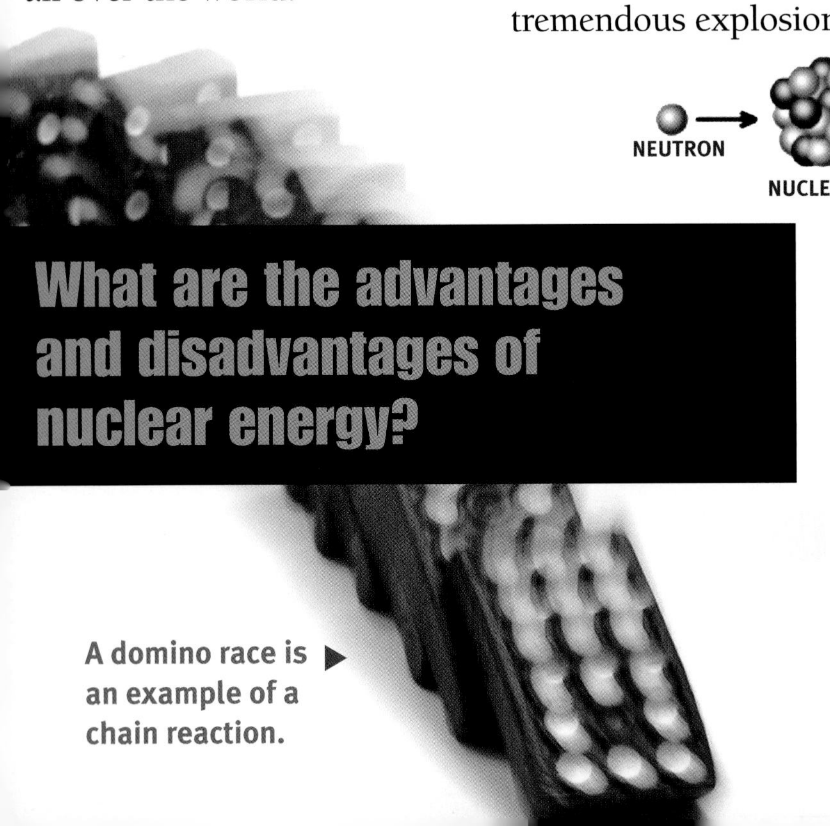

A domino race is an example of a chain reaction. ▶

Essential Vocabulary

- **fission** page 27
- **nuclear power** page 28
- **nucleus** page 26
- **radioactive** page 27

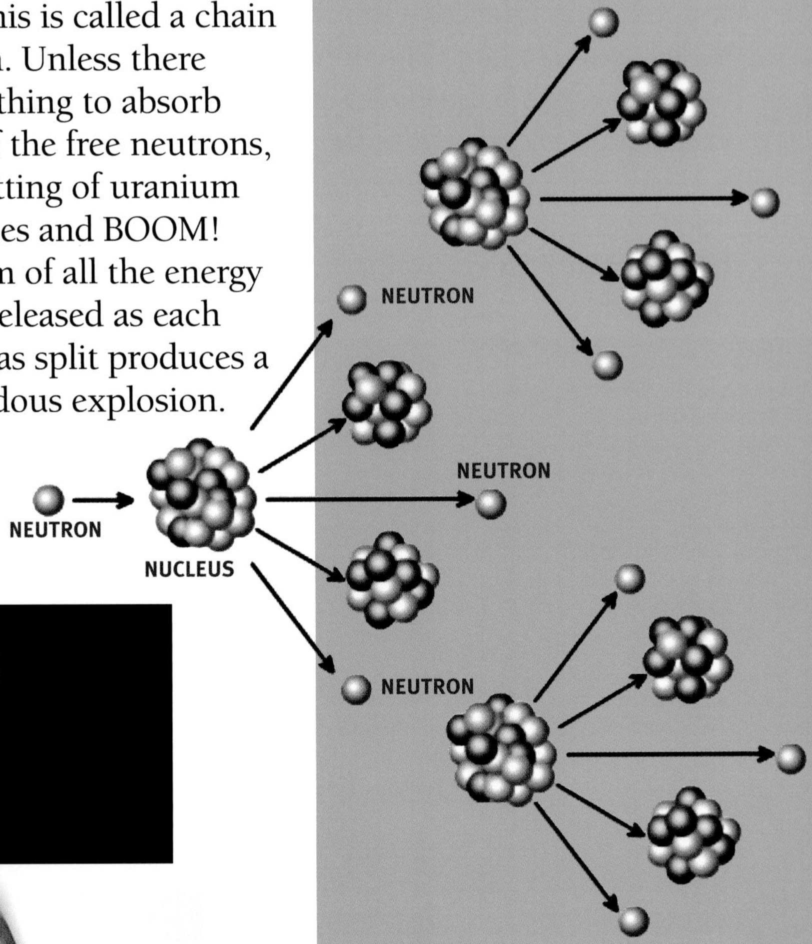

▲ As each radioactive nucleus breaks apart, it produces two or three neutrons that collide with other nuclei.

Electricity generated using nuclear energy is called **nuclear power**. Nuclear energy accounts for 2% of energy production and 15% of electricity generation in the world.

In a nuclear power plant, this energy heats water to produce steam. The steam can spin turbines to generate electricity. In the photograph below, you can see the cooled steam rising from the two cooling towers. The kinetic energy of steam can also propel nuclear submarines and warships.

A sealed building called a reactor contains the radioactive uranium fuel at a nuclear power plant. The domed building in the photograph below is the reactor. Almost no radiation escapes from the reactor. In fact, radioactive elements in coal release much more radiation into the environment than nuclear power production!

Nuclear power plants are very expensive to build. Once built, they last for decades and cost relatively little to operate. Nuclear power plants release little pollution and produce no carbon dioxide. Nuclear energy is a nonrenewable resource, but the world's uranium supply should last for centuries.

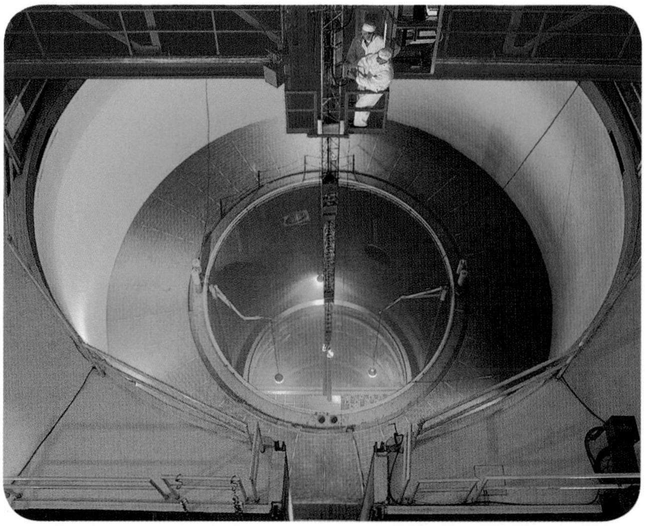

▲ A nuclear reactor emits a beautiful blue glow.

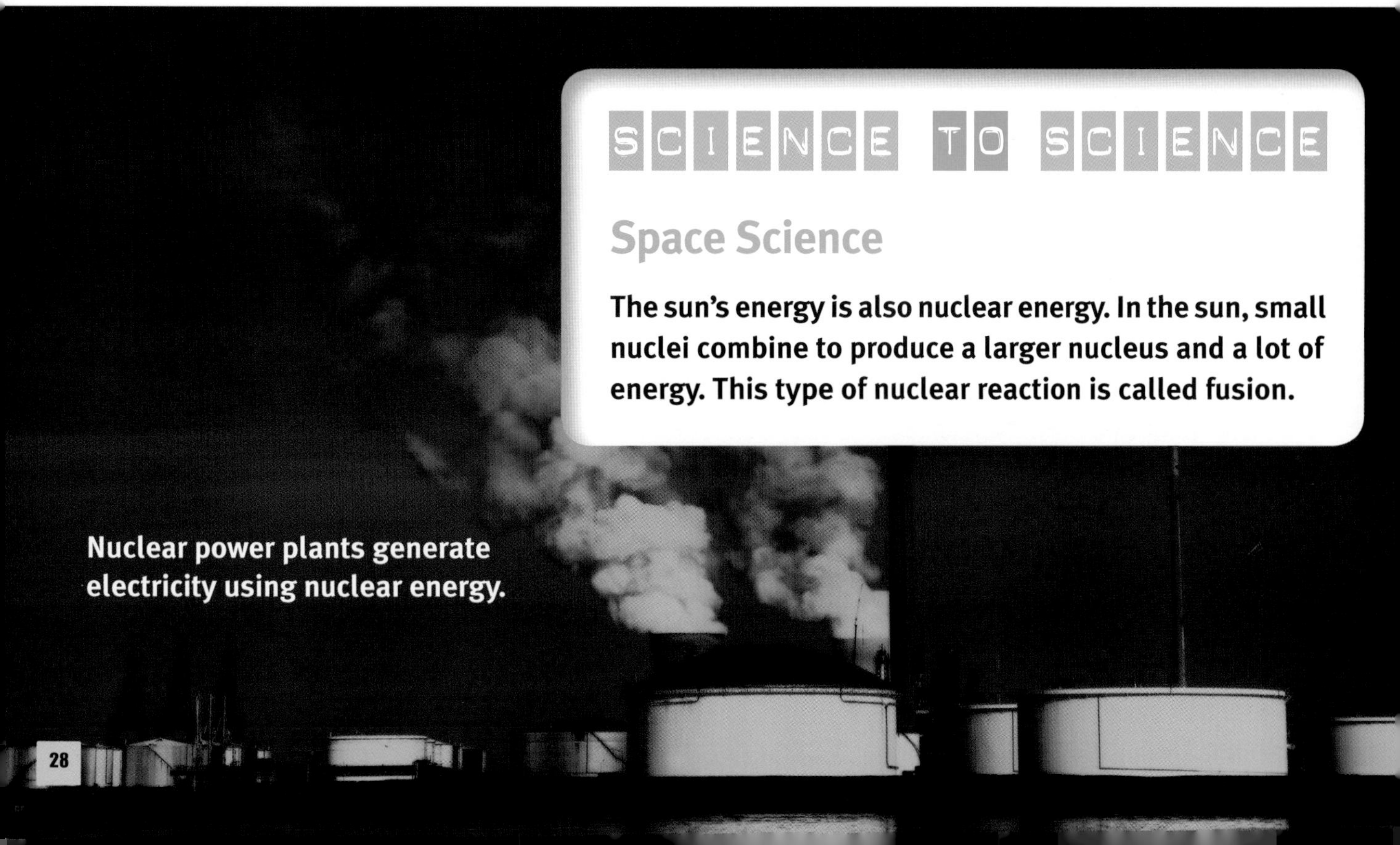

SCIENCE TO SCIENCE

Space Science

The sun's energy is also nuclear energy. In the sun, small nuclei combine to produce a larger nucleus and a lot of energy. This type of nuclear reaction is called fusion.

Nuclear power plants generate electricity using nuclear energy.

Hands-On Science

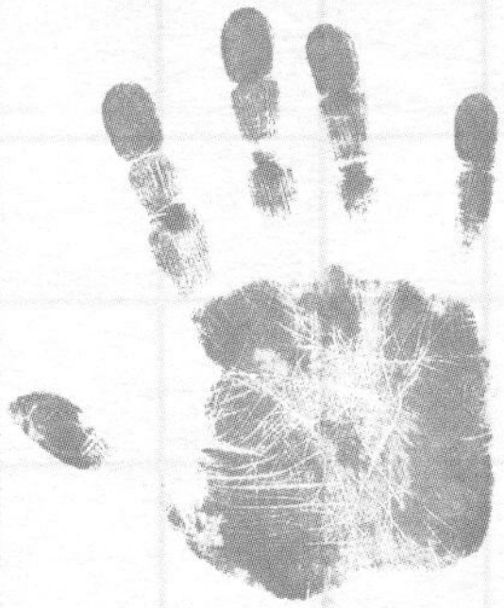

Chain Reaction

You can use dominoes to make a chain reaction. How does your chain reaction compare to a nuclear chain reaction?

TIME REQUIRED
20 minutes

MATERIALS NEEDED
- 15 dominoes
- ruler
- stopwatch or watch with second hand

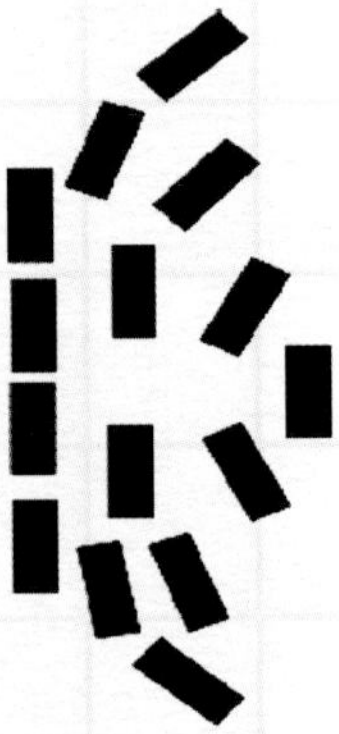

PROCEDURE

1. Stand 15 dominoes on their ends in a straight row. Set the dominoes about 2.5 cm (1 in) apart. Knock over the first domino with your finger. Time how long it takes for all the dominoes to fall. Record your results. Repeat the experiment two more times and calculate the average time.
2. Arrange the dominoes so that each domino will knock over two others. Use the illustration below as an example. Observe what happens when you knock over the first domino. Time and record how long it takes for the whole set of dominoes to fall over. Repeat the experiment two more times and calculate the average time.
3. Repeat step 2, but this time hold a ruler on its end in the middle of the arrangement. Knock over the first domino and observe what happens.

ANALYSIS

1. How is the domino race similar to a nuclear chain reaction? How are they different?
2. Compare the time it took for all the dominoes to fall in step 1 and step 2. Explain any differences you observed.
3. What effect did the ruler in step 3 have on the chain reaction?

DATA TABLE

TIME FOR ALL DOMINOES TO FALL	STEP 1	STEP 2	STEP 3
First trial			
Second trial			
Third trial			
Average time			

Safety and Waste Disposal

Every energy resource has limitations. The main concern with nuclear energy is the release of radioactive materials. Exposure to radiation can cause burns, cancer, or death.

Accidents

An uncontrolled nuclear reaction can generate enough heat to cause a meltdown. A meltdown destroys the reactor and releases radioactive material into the environment. To prevent accidents, reactors contain rods. These rods absorb neutrons and slow down the reaction. These rods act in a similar way to the ruler in step 3 of the experiment on page 29. In addition, water constantly cools the reactor.

There have been two meltdowns in over fifty years of nuclear power production. In 1979, the Three Mile Island plant in Pennsylvania melted down but caused no injuries. A more serious meltdown occurred in 1986 at the Chernobyl plant in Ukraine. About sixty people died and many local residents developed cancer.

Nuclear Waste

Used nuclear fuel and other wastes can remain dangerously radioactive for decades or even thousands of years. At this time, nuclear wastes are stored at the nuclear power plants that produce them. The federal government and the nuclear power industry are working to find other solutions. Some nuclear wastes can be processed into new fuels. Other ideas include launching the waste into space or burying it deep underground. Meanwhile, nuclear power plants continue to produce radioactive waste.

Nuclear Terrorism

Governments around the world work to keep radioactive materials secure. Nuclear fuel in the wrong hands could be used to construct a nuclear weapon or spread radioactive material around the United States.

The limitations of nuclear energy have prevented the construction of nuclear power plants in the United States since 1980. Now, however, dozens of plants are in the planning stages. The rising cost of fossil fuels and concerns about global climate change are convincing many people to give nuclear energy another chance.

Energy Resource	Advantages	Disadvantages
	• Produces no air pollution or carbon dioxide • Large supply of fuel available • Inexpensive to produce after plant is built • Excellent safety record	• Dangerous wastes last for many years • Possibility of accident or terrorist attack • Power plants very expensive to build

SUMMING UP

- Nuclear energy is produced by splitting, or fission, of atoms.
- People use nuclear energy to produce electricity and power ships.
- Nuclear power produces little pollution or carbon dioxide. On the other hand, there are concerns about the safety of nuclear power plants and the handling of their fuel and wastes.
- The United States has yet to decide the role nuclear energy will play in the future.

Putting It All Together

Choose one of the activities below.

1 Research the parts of a nuclear reactor and the purpose of each part. Make a poster to present your findings to your class.

2 Read more about the Yucca Mountain project in a library or on the Internet. Present the arguments for and against this planned nuclear waste storage facility to your class.

3 Lise Meitner (LEE-zuh MITE-ner) was the first scientist to calculate the enormous energy released by splitting an atomic nucleus. Use your library or the Internet to learn more about her interesting life. Discuss her difficult choices with a friend.

CHECKPOINT

Think About It

Even though coal production kills many more people than nuclear power production, people worry more about the dangers of nuclear power. Why do you think that is?

CHAPTER 4

Renewable Energy Resources

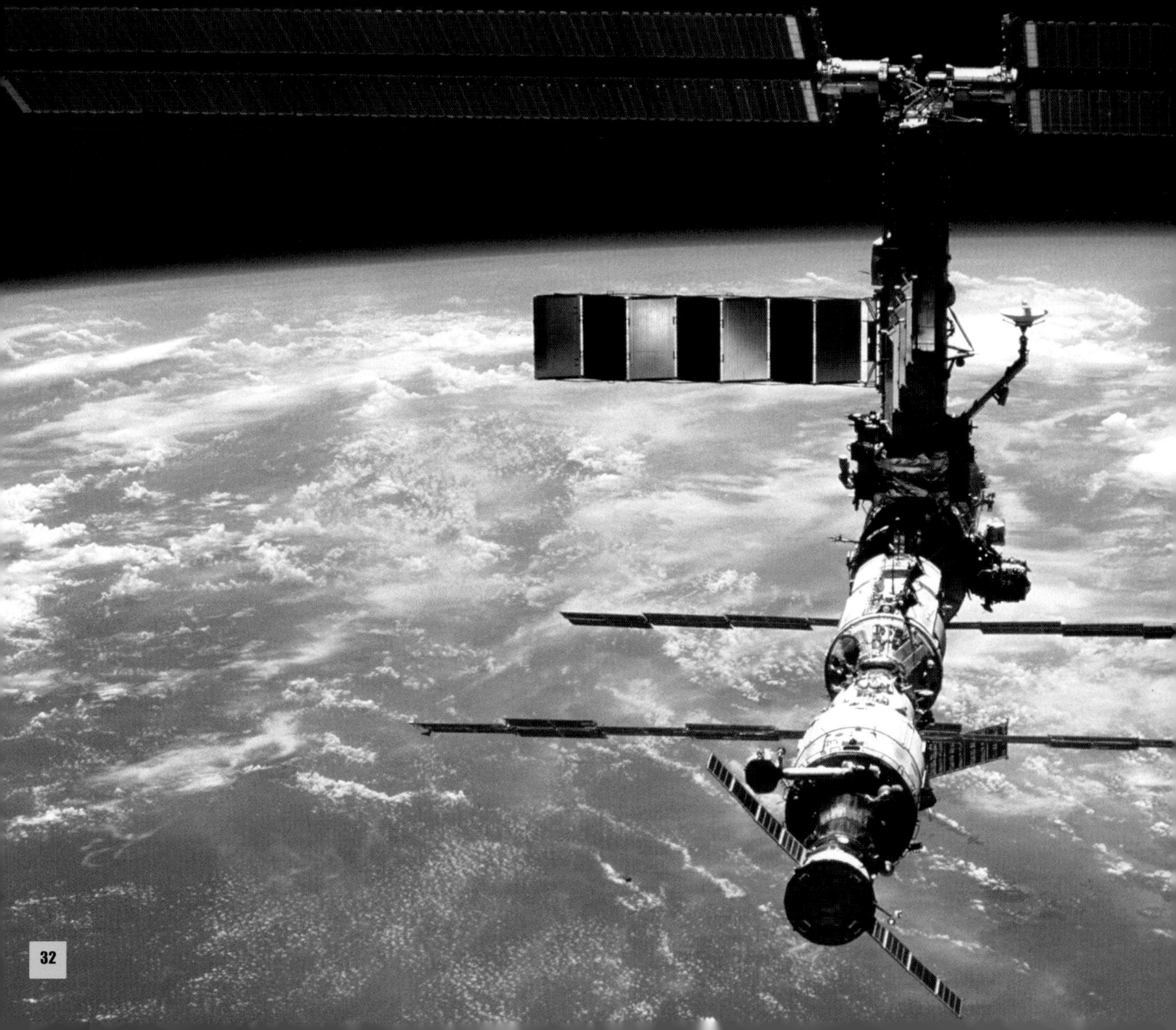

The world's supply of nuclear energy and fossil fuels shrinks a little each year. Once people consume these energy resources, they cannot make more of them. Fortunately, there are also several renewable energy resources. People have used energy from the sun, wind, moving water, biomass, and Earth for thousands of years. Modern technology provides new ways to take advantage of these renewable resources.

Essential Vocabulary

- biomass energy — page 39
- geothermal energy — page 38
- hydroelectric power — page 36

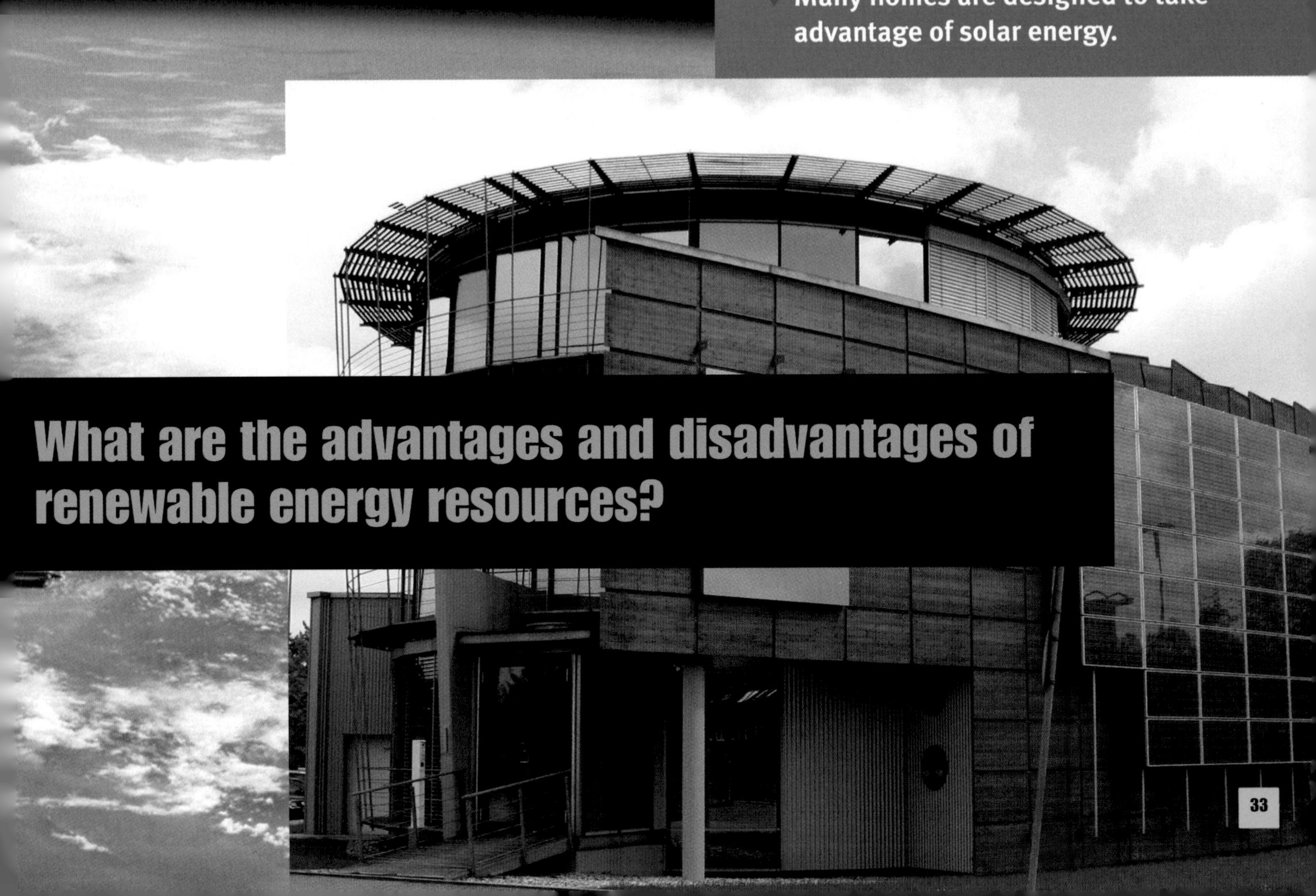

Many homes are designed to take advantage of solar energy.

What are the advantages and disadvantages of renewable energy resources?

Solar Energy

Energy from the sun was one of the first energy resources used by people. Ancient people built thick-walled homes that absorbed sunlight during the day and radiated that heat back into the home during the cold night. People also used sunlight to heat water and dry food. These passive solar energy applications do not require mechanical systems.

An enormous amount of sunlight reaches Earth's surface. Twenty days of solar energy contain as much energy as all of the reserves on Earth. Active solar energy technology uses collectors and other devices to convert solar energy into useful heat, light, and electricity.

CHECKPOINT

Make Connections

Think about how the sun influences your life. Could life go on without the sun?

▼ **Percentage of energy produced by each energy source in the United States**

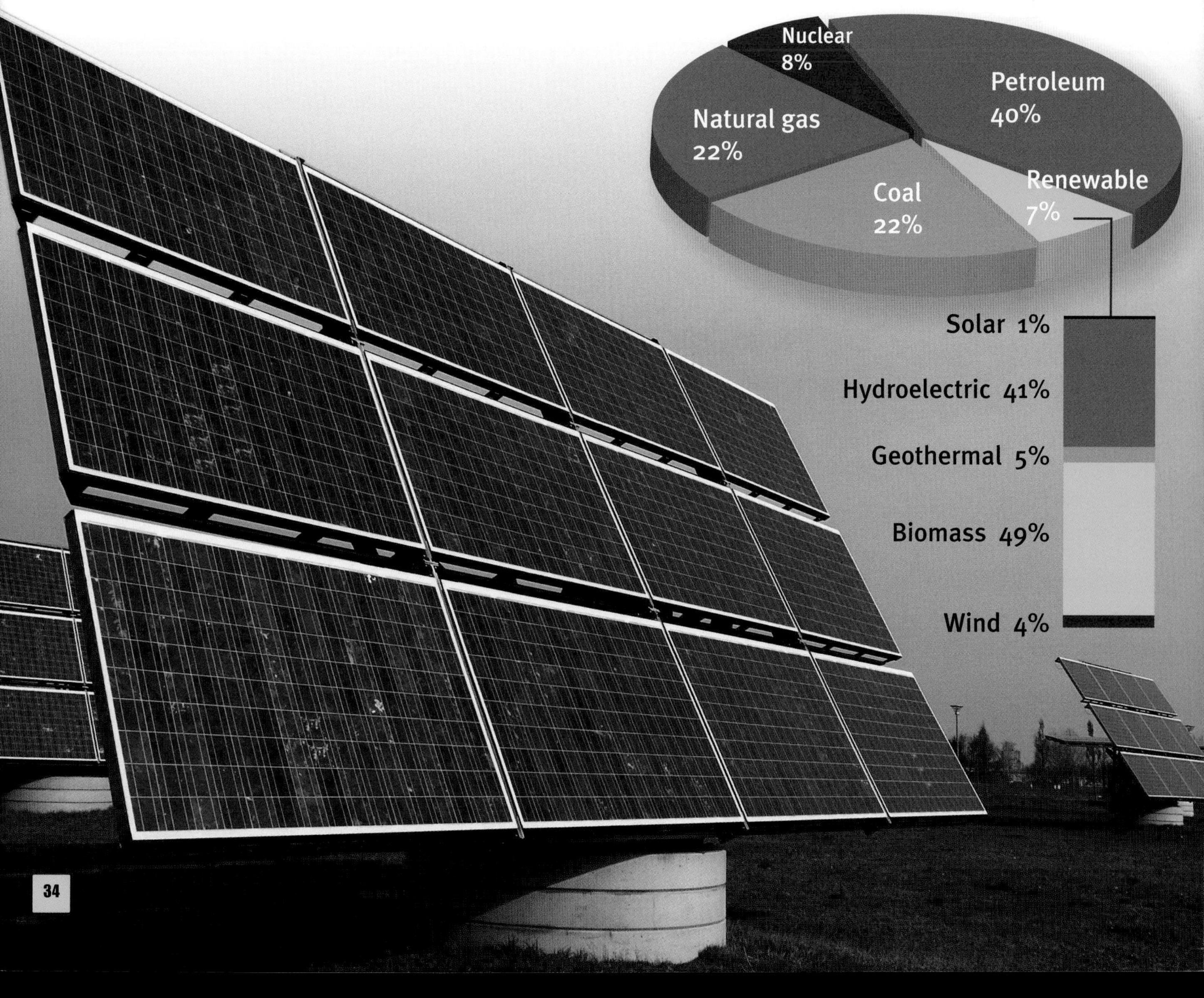

Have you ever seen solar panels on a roof or outdoor lighting? Solar panels contain solar cells that collect sunlight and generate electricity. Electricity from solar energy is called solar power. In the United States, California, Nevada, and Arizona all have solar power plants. So far, solar cells are expensive and inefficient—they convert only a small percentage of sunlight to electricity. Sunlight is only available part of the day. Solar energy must be stored in batteries for use when the weather is not sunny.

For these reasons, solar power costs more today than electricity generated from burning fossil fuels. However, researchers constantly improve solar technology to bring down prices. Solar panels already cost less than laying power lines to remote locations. Solar power also makes sense in satellites and space vehicles.

Advantages and Disadvantages of Solar Energy

ENERGY RESOURCE	ADVANTAGES	DISADVANTAGES
	• Produces no air pollution or carbon dioxide • Enormous supply • Renewable	• Only available on sunny days • Solar cells inefficient and expensive • Amount of sunlight is not constant at a given location • Requires large surface areas to collect the energy to be used at a useful rate

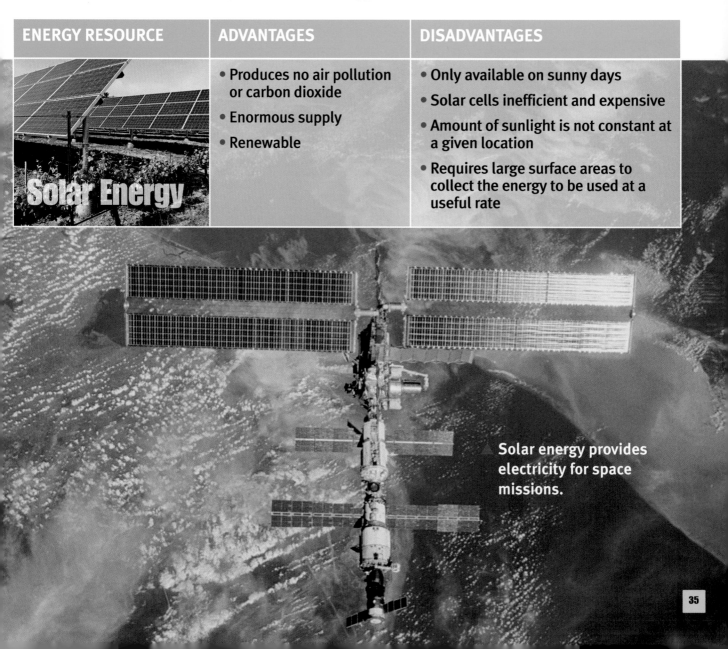

Solar energy provides electricity for space missions.

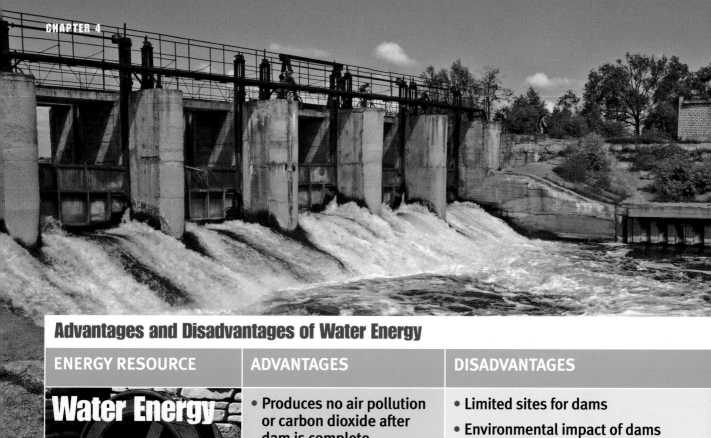

Advantages and Disadvantages of Water Energy

ENERGY RESOURCE	ADVANTAGES	DISADVANTAGES
 Water Energy	• Produces no air pollution or carbon dioxide after dam is complete • Relatively inexpensive electricity • Enormous supply • Renewable	• Limited sites for dams • Environmental impact of dams and flooding • Building dams releases carbon dioxide • Can disrupt plant and animal habitats • Sometimes can force relocation of people

Water Energy

Flowing water is an example of kinetic energy. People harnessed the energy of moving water to turn wheels and run mills over 2,000 years ago. Now, dams collect water and use it to spin turbines that generate electricity, or **hydroelectric power**. Moving water costs nothing and creates no pollution or carbon dioxide.

Why don't we get all of our energy from moving water? First, there are not enough rivers to dam. Engineers are working to get around this limitation. They are developing hydroelectric power plants that use the energy of ocean waves and tides.

Second, dams and their construction impact the environment. The making of concrete for dam construction releases a lot of carbon dioxide. This carbon dioxide contributes to global climate change. In addition, dams flood the areas behind them and reduce water flow downstream. States have begun to destroy some hydroelectric dams because they cause harm to local fish populations.

Wind Energy

Moving air is also an example of kinetic energy. Early uses of wind energy included sailing ships and using windmills to pump water and grind grain. The wind's kinetic energy now spins turbines and generates electricity. Turbines that generate wind power can spread 126 meters (413 feet) across. Large wind farms produce enough electricity to power whole towns while farming continues under the turbines. Wind power produces no pollution or carbon dioxide. In addition, it costs less to build wind turbines than hydroelectric dams or nuclear power plants.

What are the limitations of wind energy? Some people object to the appearance of large wind turbines, which can be seen for many miles. Studies show that migrating birds and bats can die from collisions with turbines or the high-pressure air they create. The single greatest limitation of wind power is the unpredictable nature of wind. Wind power works best when combined with other energy sources.

CAREERS IN SCIENCE

Renewable-Energy Scientist

Renewable-energy scientists work to improve the use of renewable energy resources. These scientists study geology, physical science, or engineering. Energy scientists work for businesses, universities, and governments to find cleaner and cheaper ways to use energy from renewable resources.

Advantages and Disadvantages of Wind Energy

ENERGY RESOURCE	ADVANTAGES	DISADVANTAGES
Wind Energy 	• Produces no air pollution or carbon dioxide • Enormous supply • Renewable • Wind turbines can share land with other uses	• Wind not always steady or predictable • Wind turbines can harm wildlife • Some people find wind turbines unattractive

Geothermal Energy

It is very hot inside Earth. This **geothermal energy** comes from the molten rock below Earth's surface. It also comes from the decay of radioactive particles. In some places, the hot rocks lie close to Earth's surface. People have used geothermal energy to heat bathing pools and homes for many years.

Power plants use geothermal energy to boil water and generate electricity. First, water is pumped underground. Heat energy from the hot rocks turns the water into steam. Steam turns a turbine to make electricity. Most geothermal reservoirs are deep underground. Gases released from Earth's depths are the only pollution resulting from this type of power production.

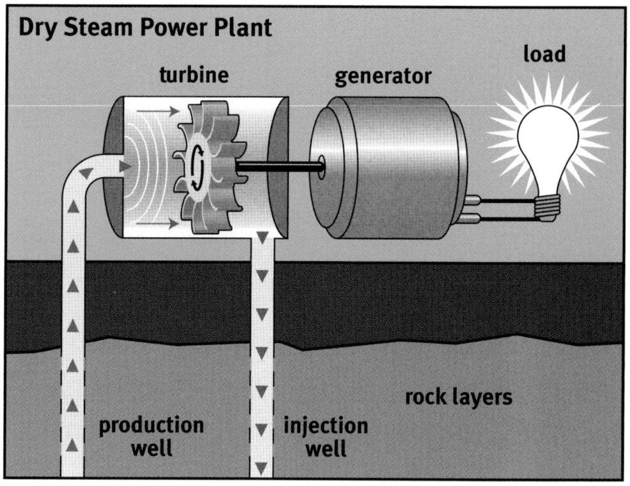

Geothermal energy is only available in certain locations. The country of Iceland generates 27% of its electricity from geothermal energy. Iceland also uses geothermal energy for most of its heating needs. Geothermal power plants in California, Hawaii, Nevada, and Utah provide electricity to millions of homes in the United States.

Advantages and Disadvantages of Geothermal Energy

ENERGY RESOURCE	ADVANTAGES	DISADVANTAGE
Geothermal Energy 	• Releases very little pollution or carbon dioxide • Enormous supply • Renewable	• Only available where hot rocks lie close to Earth's surface

Geothermal power plants use Earth's heat to generate electricity.

Biomass Energy

The Biomass Fuel Cycle

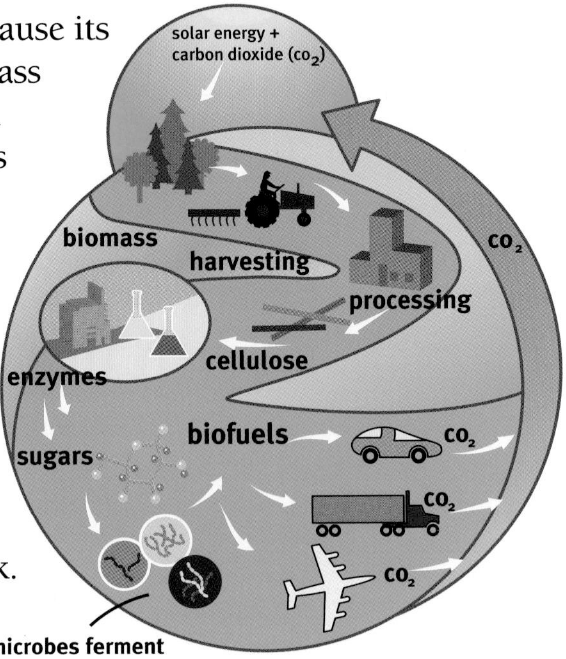

Biomass energy is the potential energy in plant material and animal waste. Biomass is a renewable resource because its energy comes from the sun. The simplest uses of biomass energy include burning wood, plant stalks, and animal wastes. The energy released by burning heats buildings and generates electricity. Biomass is also converted into biofuels such as ethanol and diesel.

Plants use carbon dioxide from the air to grow. Burning biomass or biomass fuel releases that carbon dioxide back into the air. The use of biomass energy produces no new carbon dioxide.

Biomass fuels are often made from food crops such as corn, soybeans, and sugarcane. Biofuel production could reduce the supply of food for people and livestock. Other sources of biomass are inexpensive or free. However, it is difficult to convert grass clippings or cornstalks into biofuels. Many scientists are working on this problem.

Advantages and Disadvantages of Biomass Energy

ENERGY RESOURCE	ADVANTAGES	DISADVANTAGES
Biomass Energy	• Fuels made from waste products are inexpensive • Produces no new carbon dioxide	• Most plant material is difficult to convert to fuel • Competition for land between food and fuel uses

SCIENCE AND TECHNOLOGY

Genetic Engineering

Plants contain tough fibers. Grass-eating animals such as cows use bacteria to break down plant fibers into sugars. Scientists are using genetic engineering to make these bacteria useful to people. The engineered bacteria could digest grass and wood into sugars for biomass fuel production.

They Made a Difference

Amory Lovins (1947 –)

Amory Lovins is an American physicist who studies energy resources. He believes that the wise use of energy resources is very important. Lovins argues that business and the environment both benefit when people reduce energy use. His ultimate goal is sustainable energy use—energy use that does not have a negative impact on the environment or the people of the future. It will take renewable energy resources to achieve sustainable energy use.

Lovins founded the Rocky Mountain Institute (RMI) to help companies and governments make sensible energy choices. RMI helped the Empire State Building in New York City reduce its energy use by nearly 40%. Lovins is also working to develop a vehicle that gets 100 miles per gallon of gasoline. Lovins is famous for finding solutions that reduce energy use and increase profits at the same time.

Amory Lovins ▶ works with businesses to reduce their energy needs.

"Our energy future is choice, not fate." – Amory Lovins

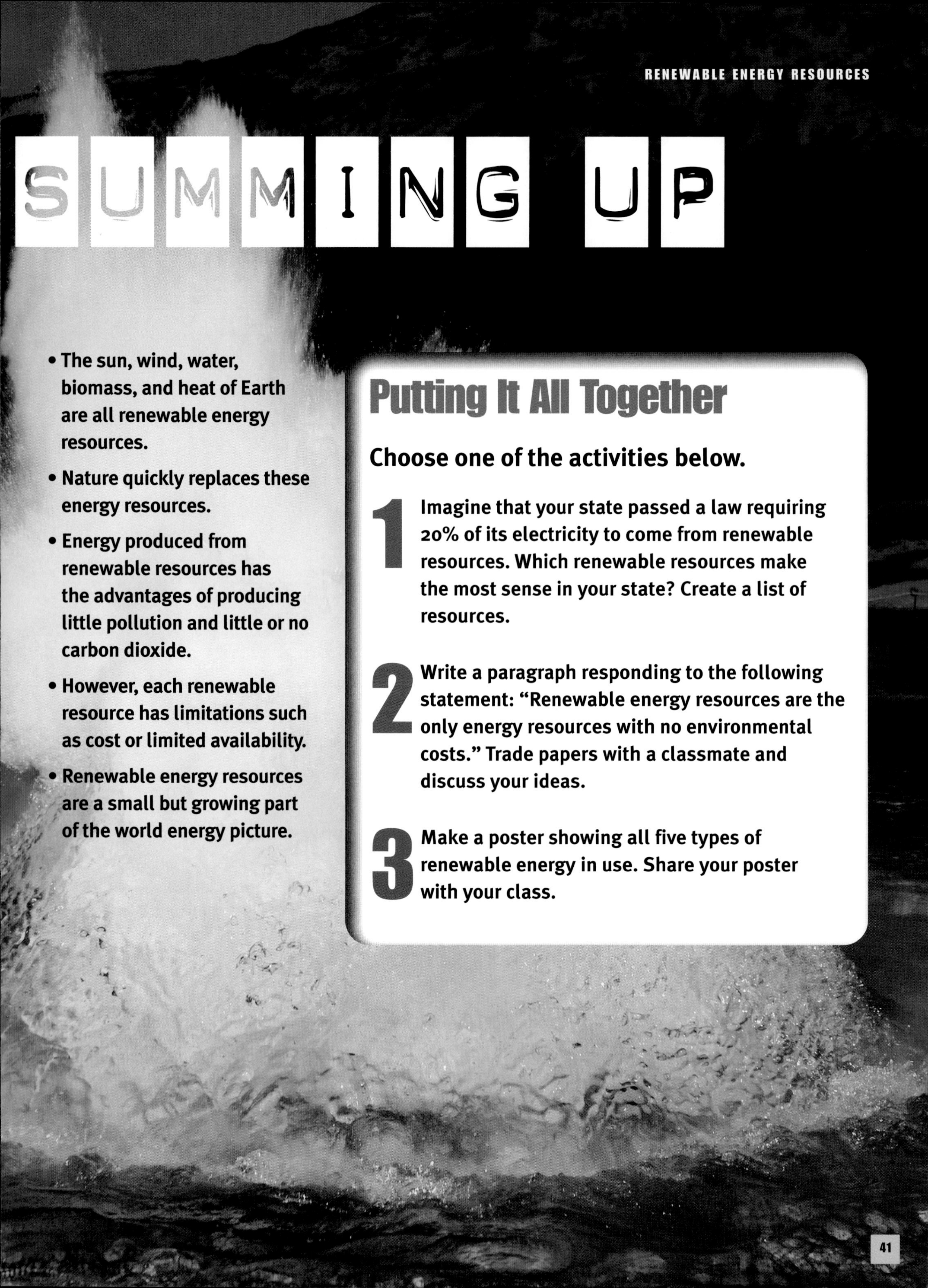

SUMMING UP

- The sun, wind, water, biomass, and heat of Earth are all renewable energy resources.
- Nature quickly replaces these energy resources.
- Energy produced from renewable resources has the advantages of producing little pollution and little or no carbon dioxide.
- However, each renewable resource has limitations such as cost or limited availability.
- Renewable energy resources are a small but growing part of the world energy picture.

Putting It All Together

Choose one of the activities below.

1. Imagine that your state passed a law requiring 20% of its electricity to come from renewable resources. Which renewable resources make the most sense in your state? Create a list of resources.

2. Write a paragraph responding to the following statement: "Renewable energy resources are the only energy resources with no environmental costs." Trade papers with a classmate and discuss your ideas.

3. Make a poster showing all five types of renewable energy in use. Share your poster with your class.

CONCLUSION

Quest for Energy

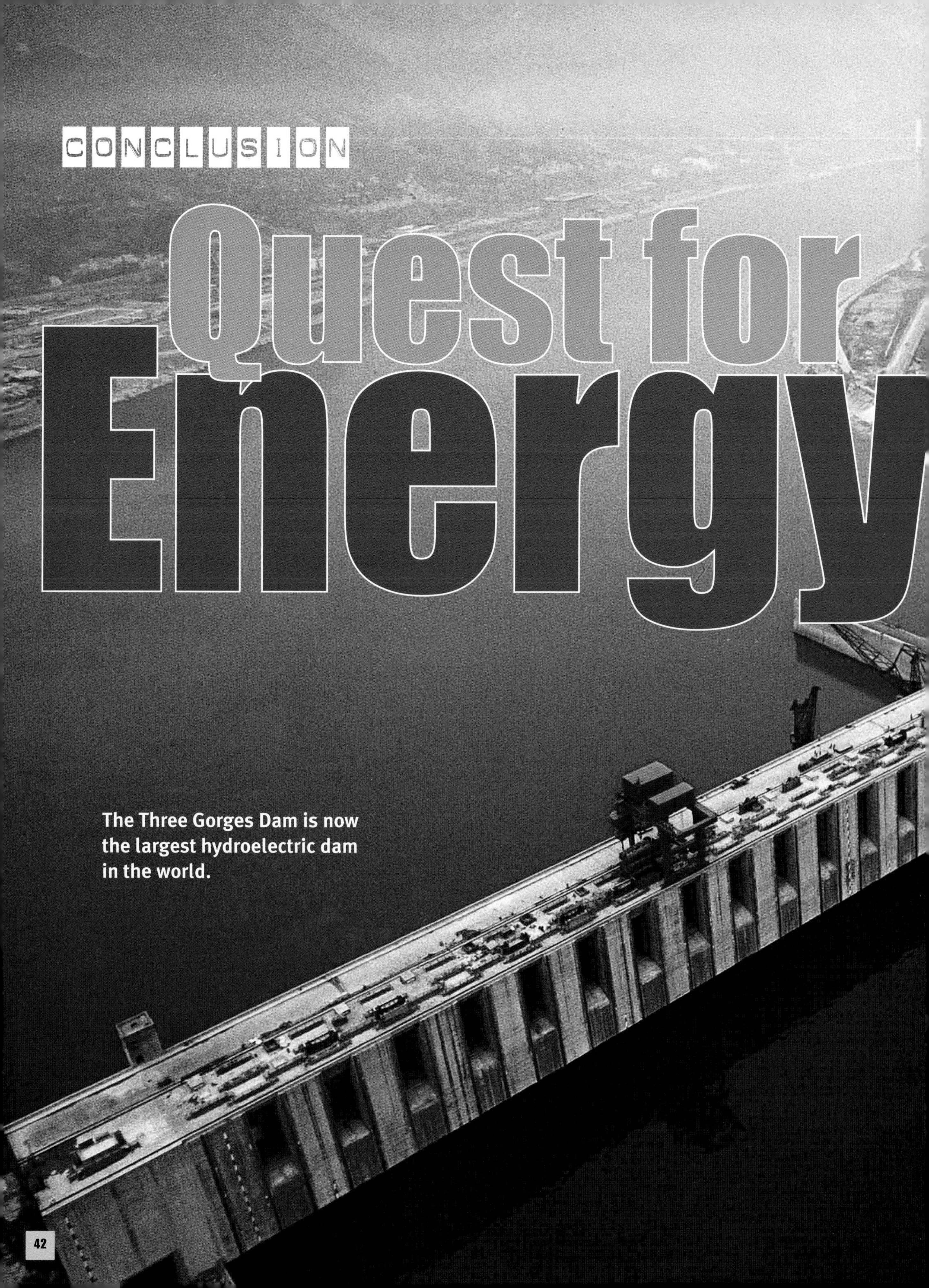

The Three Gorges Dam is now the largest hydroelectric dam in the world.

THE CHINESE GOVERNMENT BEGAN BUILDING THE THREE GORGES PROJECT IN 1994. In 2003, the hydroelectric power plant began to produce electricity. The dam now produces more electricity than a dozen nuclear power plants. It also replaces the burning of forty to fifty million tons of coal each year. The sale of electricity will pay for dam construction by 2017.

The Three Gorges Dam also created many problems. The rising waters behind the dam displaced millions of residents. There were also major environmental effects. Making the concrete for the dam released a tremendous amount of carbon dioxide into the atmosphere. The change to the river also resulted in the extinction of the Chinese river dolphin, riverbank collapse, and droughts.

How will we power our lives in the future? No one energy resource holds all the answers—we will need a mix of several. Scientists and engineers are developing ways to use every energy resource in cleaner, cheaper, and safer ways. Reducing energy use is the cleanest and safest improvement of all.

Buildings in the future may use multiple energy resources along with energy-saving features.

How to Write a Persuasive Letter

Have you ever felt so passionate about a subject that you wanted to let everyone in your community know about it? You can spread the word with a letter to the editor of your school or local newspaper. The newspaper may print your persuasive letter on the editorial page or the newspaper's Web site.

STEP 1 Choose a topic that is important to you and to your community. For example, you might want solar-powered lighting installed in a new skate park.

STEP 2 Carefully read the guidelines for letters to the editor. You can find these guidelines on the newspaper's Web site or editorial page.

STEP 3 Collect facts and figures that you want to present. Make an outline of your most important points.

STEP 4 Write your letter as clearly and briefly as possible. State your main point in the first sentence. Follow your main point with a sentence or two giving background information. End by suggesting what the reader can do to help.

STEP 5 Type your letter and proofread it carefully. Keep your tone respectful. A well-written, polite letter will convince more readers than an angry, careless note.

STEP 6 Sign your name and give any required contact information. Submit your letter by mail or e-mail.

Sample Letter

To the Editor:

Deerwood's new skate park is missing one important detail: lighting. I believe that the park would be much safer with lights around the edges. Without lighting, accidents will be more likely between dusk and the park closing at 8 P.M.

Solar-powered lighting is ideal for the skate park. The lights are only needed for a couple of hours each day. Solar lights do not require any power lines or extension cords. During the day, the lights store energy from sunlight. In the evening, the lights come on automatically and shine for a few hours. Solar lamp posts cost about the same as traditional street lights. After purchase, solar lights cost nothing to operate.

Deerwood residents can contact their council member about adding lights to the skate park. Together, we can save money and keep skaters safe.

Sincerely yours,

Freddie Marzano

Glossary

biomass energy (BY-oh-mas EH-ner-jee) *noun* potential energy found in plant material and animal waste (page 39)

energy (EH-ner-jee) *noun* the ability to do work (page 7)

energy resource (EH-ner-jee REE-sors) *noun* an energy source used to meet the needs of people (page 8)

fission (FIH-shun) *noun* the splitting of one atomic nucleus into two nuclei with the release of a lot of energy (page 27)

fossil fuel (FAH-sul FYOOL) *noun* an energy resource that is formed inside Earth from ancient plant or animal remains (page 16)

geothermal energy (jee-oh-THER-mul EH-ner-jee) *noun* energy from inside Earth (page 38)

hydroelectric power (HY-droh-ih-LEK-trik POW-er) *noun* electricity generated from the energy of moving water (page 36)

kinetic energy (kih-NEH-tik EH-ner-jee) *noun* energy of movement (page 7)

nonrenewable resource (nahn-rih-NOO-uh-bul REE-sors) *noun* a natural resource that cannot be replaced (page 8)

nuclear power (NOO-klee-er POW-er) *noun* electricity generated using nuclear energy (page 28)

nucleus (NOO-klee-us) *noun* the core of an atom containing protons and neutrons (page 26)

potential energy (puh-TEN-shul EH-ner-jee) *noun* stored energy (page 7)

radioactive (ray-dee-oh-AK-tiv) *adjective* spontaneously releasing energy and particles from an atomic nucleus (page 27)

renewable resource (rih-NOO-uh-bul REE-sors) *noun* a resource quickly replaced by nature (page 8)

Index